SCIENCE

IN EVERYDAY THINGS

SCIENCE
IN EVERYDAY THINGS

by

WILLIAM C. VERGARA

Illustrated

HARPER & BROTHERS, PUBLISHERS, NEW YORK

Library of Congress catalog card number: 57-11790

TO

PATRICIA AND BOB

PREFACE

Most of us are curious about science. We would like to share in the excitement that seems to surround it. Yet, I've heard many intelligent people comment, "I'll never be able to understand science— it's too technical!" I suspect that many of these people have had an unfortunate early encounter with a disinterested teacher or a dull textbook. This can frustrate even the most intelligent and interested student. But science doesn't have to be dull. There are many well-written and interesting scientific books on the market. The problem is finding them. They are often well hidden among shelves of textbooks and treatises. The former are usually written for teachers and the latter for scientists. In *Science in Everyday Things* I have tried to explain the wonders of science to the nontechnical reader. In so doing, I have tried to give science a fighting chance. You will find that only the essential points are considered. Mathematical formulas, charts and experimental data are studiously avoided. We won't need them. On the other hand, no unnecessary piefilling has been added; every sentence has been made to pay its own way.

Many of the questions and answers are simple; a few are difficult. In most cases, the underlying scientific principle is presented as well as the answer. Most people are not satisfied with a simple yes or no; they insist on knowing "How come?" Where science is scratching its puzzled head, I have so stated—giving the most educated guesses available. I've tested most of the answers on my long-suffering, nontechnical friends and where blank stares have resulted, I have retreated to the typewriter for a renewed assault on clarity. I hope my efforts have met with some success. I have tried to talk slowly and clearly on what science is all about.

I've included questions on just about every branch of science, probably to satisfy my own curiosity. The answers have come from many current books and periodicals. The absence of footnotes and references is certainly not intended to connote originality. I am indebted to hundreds of authors for the information that I have put together. I regret that their numbers makes it impracticable to acknowledge that debt. Here and there I have thrown in an extra bit of information, just for good measure. These asides seemed interesting to me, yet not important enough to warrant a question of their own. Their addition seemed to be in keeping with the spirit of the book.

To the casual observer, it may appear that the questions in this book are located without the slightest thought to logical order. He couldn't be closer to the truth. First, I separated the questions into haphazard groups; then I shuffled them; and then I put them back together at random. This was done to add to the interest of the general reader, and to avoid the pattern of the textbook. I have tried to make the book a treasure trove of interesting though seemingly unrelated scientific information. The index is its key.

In writing this book, I have tried to deal in facts. I believe these facts to be true. But science has an annoying habit of changing its mind. That which is accepted today as fact may be carelessly discarded tomorrow as fancy. No less a scientific personage than Sir Isaac Newton has had his theory of light pass through the stages of acceptance, rejection and cautious reappraisal. No one can say, "This is a scientific fact!" But each of us can make use of scientific information that seems to work today.

I have received much valued advice and help from many of my friends and colleagues—to the advantage of the book. I am deeply grateful. To list their names here would be impracticable. But I can't pass by the opportunity to mention the name of a man who has given so freely of his kindly encouragement and indispensable editorial counsel—my brother, Joseph R. Vergara. If this book should be found helpful, it will be in greatest measure due to his interest and advice.

WILLIAM C. VERGARA

Phoenix, Maryland
July, 1957

SCIENCE

IN EVERYDAY THINGS

SCIENCE
IN EVERYDAY THINGS

⊂╪

Can rocket ships fly in a vacuum?
Sir Isaac Newton provides a clue to the answer to this question in his third law of motion—"To every action there must be an equal and opposite reaction." To illustrate, imagine that you and a companion, both wearing ice skates, are standing face to face on the ice. In addition, all four skates are parallel to each other. What would happen if you now gave your friend a gentle shove? Both of you would move in opposite directions from the starting point. He would move in response to the force of your push, and you would move in response to the equal and opposite force exerted by his body on your hands.

Now let us assume that you are sitting in a stationary canoe on a quiet pond. If you were to pick up a heavy object and thrust it far over the side you would note the canoe beginning to move in the opposite direction. Your hands, in exerting a force on the weight, would, in return, be acted upon by the weight. This reaction causes the canoe to move in opposition to the direction of the toss. Now suppose your canoe is full of lead balls. Why not propel yourself along by throwing the balls away over the stern of the canoe? If you had two or three helpers throwing balls, the canoe would move even faster. As a matter of fact, the speed of the canoe would depend on how hard and how quickly you threw lead balls away.

Rocket engines operate on exactly the same principle. The only difference is the substitution of gas particles for the lead balls. The engine pushes gas particles out the rear of the rocket at very high speed. The reaction pushes the rocket forward. Since the engine does not depend on our atmosphere for its operation, it will work in a vacuum. As a matter of fact, its performance in a vacuum would be superior since air friction would be absent and could not provide opposition to the rocket's movement.

FIG. 1.   A ROCKET ENGINE USING A SOLID PROPELLANT
As the fuel burns, hot gases are forced out of the opening at the rear of the rocket. The reaction pushes the rocket forward.

What are the "bends" that sometimes afflict divers?

This condition results from the fact that the amount of gas which can dissolve in a liquid goes up as the pressure increases. Examine an unopened bottle of ginger ale or soda pop. The liquid is clear and no gas bubbles are evident. When we remove the cap from the bottle, carbon dioxide gas rushes out and bubbles immediately begin to form in the liquid and escape to the surface. Since gas rushes out, the liquid must have been under a pressure greater than atmospheric before the bottle was opened. When the pressure on the liquid was reduced, it was no longer capable of containing as much gas in solution. As a result, the excess gas came out of solution, forming bubbles which were released to the atmosphere.

The same sort of thing happens in man. If a man is placed in an environment where all of the gas surrounding him is under high pressure, the fluids of his body dissolve an abnormally high amount of gas. This fact is not harmful in itself, but if this pressure is suddenly released, gas bubbles form in his body fluids just as they did in the opened bottle of ginger ale. Some of these bubbles may be large enough to block blood vessels and interfere with the blood's natural circulation. Frequently, they affect organs in the middle of the body, causing the individual to bend over in pain. It is this effect which gave the name "bends" to this condition. It is most often observed when deep-sea divers are elevated too rapidly from the depths. Its less common name is *decompression sickness*.

2

What is the life history of a star?

There has been a great deal of speculation concerning the origin, development and decline of stars. We know that they differ tremendously in temperature, size and density and to some extent in the material they are made of. The red supergiants at one extreme are many times larger than our sun while the white dwarfs resemble our own earth in size. The supergiants have a density only one-thousandth as great as the air we breathe while the white dwarfs are hundreds of thousands of times as dense. A current theory is based on the assumption that all stars must go through both of these phases in their life history. The theory states that cosmic dust clouds contract due to attraction between the particles, eventually becoming gaseous and beginning to glow as red supergiant stars. Further contraction brings such stars to the size and temperature of average stars like our sun. It is believed that the period of youth as a supergiant is relatively short and this belief is substantiated by the scarcity of such stars in the heavens. After becoming average stars they settle down for many billions of years as stable members of the celestial community. During this period they radiate energy obtained by changing hydrogen to heavier elements. Our sun is now in this period of its life cycle. When the supply of hydrogen is almost exhausted, the stars collapse and undergo several severe explosions, following which they regain stability as very dense white dwarfs. Like that of youth, the period of old age of stars is relatively short. With dwindling sources of energy, the star soon begins to lose its brightness and finally ceases to shine altogether.

Why do geysers spout?

A geyser is a hot spring which erupts intermittently, sending a column of boiling water and steam high into the air. Geysers are found in New Zealand, Iceland and in our own Yellowstone National Park. They usually have a tube leading down to hot rock, a basin or bowl around the opening, and a near-by supply of underground water to fill the tube.

The present-day explanation of geysers was first proposed by Bunsen as a result of his observations made on the geysers of Iceland. Bunsen knew that the temperature at which water boils depends upon the pressure exerted upon the water. This temperature is 100° C. at sea level (pressure, 14.7 pounds per square inch) and increases gradually as the pressure increases. If one were to fill a vertical tube with 300 feet of water, he would find the pressure at the bottom of the tube equal to about 10 times atmospheric pressure or 147 pounds per square inch. At this pressure, water must be raised to about 180° C. before it begins to boil.

A geyser is likely to occur when such a tube exists in the earth in the vicinity of water and hot rock. Water seeps into the tube and becomes heated by contact with the hot rock. The temperature of this water increases gradually until steam is formed in spite of the tremendous pressure. This steam then forces the column of water to shoot into the air in much the same manner as a common percolator coffee pot. When the superheated water hits the reduced pressure of the atmosphere, some of it bursts into steam causing a spectacular eruption. The regularity with which many geysers erupt indicates that the time cycles for the various operations have become extremely stable.

Why do plants turn toward light?
Plants don't really turn at all but, in reality, grow toward the direction of light. This characteristic is due to the accumulation on the dark side of plants of unusually large amounts of growth-stimulating hormones. Just as in animals, these hormones, or *auxins* as they are called, help to regulate the chemical activity going on in the plant. They are generally secreted in one part of the plant and perform their functions in entirely different parts. They stimulate growth and development and seem to control the activities of certain tissues and organs.

Auxins are the best-known family of plant hormones. It is now generally believed that the growth of plant cells will take place only in the presence of minute amounts of auxin. Auxins are produced mainly in plant tissues of young developing leaves and buds. From these points, they are transported down to other parts of the plant where they stimulate growth.

For some unknown reason the auxin circulates in greater quantity into those cells located on the darker or shaded portions of the plant. This means that the side of the plant located away from the sun or other source of light will receive more than its share of the auxin. As a result, the cells on the dark side of the stem elongate faster than those on the light side of the stem. The longer cells on

4

the dark side naturally bend the stem in the direction of the light. As a result, plants seem to seek the source of light.

What is the shape of a raindrop?

Most of us have seen drawings depicting raindrops with their gracefully streamlined teardrop form. Unfortunately, such is not the case. Raindrops are actually almost spherical in shape. When a glass of water is spilled on the floor, the force of gravity tends to spread the water out in all directions since water molecules roll over each other with ease. A falling raindrop, however, is not subject to these distorting forces and tends to assume the most convenient shape. This is due to the fact that surface films of liquids are *elastic*. A needle or razor blade can be floated on water because of this elasticity. If you blow a soap bubble and then remove the pipe from your mouth the bubble slowly contracts due to this elasticity. It is only natural, therefore, for free liquids, such as raindrops, to assume the type of configuration which has the smallest possible surface area. Geometry tells us that a sphere has less surface area for its volume than any other geometrical form. As surface forces act on the raindrop, they tend to reduce the surface area. This pulls the liquid into the shape of a sphere.

Why does penicillin cure disease?

Penicillin is an organic compound consisting of carbon, hydrogen, oxygen, sulfur and nitrogen. It does not kill the germs against which it is used, but merely stops their growth and prevents their reproduction. Penicillin and other organic compounds such as streptomycin and aureomycin are called *antibiotics*. Antibiotics are organic compounds made by living organisms and effective in stopping the growth of certain specific germs.

It has long been known that certain microorganisms have the ability to inhibit the growth of bacteria. In 1928 Dr. Alexander Fleming, while experimenting with bacteria, noticed that a colony of the mold *Penicillium notatum* had accidentally infected the culture. This mold seemed to prevent the growth of the bacteria over a considerable area, an area greater than that over which the mold was growing. It was evident that it was not the mold itself that affected the bacteria, but some material produced by the mold. No attempt was made to use this substance, however, or to identify similar ones until 1939 when Dubos of the Rockefeller Institute isolated tyrothricin and showed that it had the ability to cure certain bacterial infections in man. The next year, the English scientists Florey and Chain reported that partly purified penicillin which they had prepared was effective against a variety of bacteria and had little toxic effect on the patient. During World War II, researchers in the United States and the United Kingdom joined their efforts to develop methods of purifying penicillin economically on a large scale. In late 1946, penicillin was finally made synthetically in the laboratory.

Since the extraction of penicillin from the mold plant in 1940, research has gone forward rapidly. We now know the chemical formula for penicillin and, in addition, we know which disease germs are stopped by it. If a drug as amazing as penicillin could be discovered by accident, perhaps there are more of even greater value still to be discovered. You may be sure that scientists are looking for them.

The reason for the effectiveness of the antibiotics is not too clear. It seems probable that disease germs are tricked into absorbing these substances because of their resemblance to the germs' normal type of food. It appears that antibiotics are enough like these food substances that bacteria are willing to absorb them, but different enough that they do not function as food. Thus the life processes of the microorganism are prevented from working properly.

What's wrong with our calendar?

I think you will agree that our present calendar is cumbersome and unwieldy. It evolved more or less by trial and error during the last 2,000 years and is not suited to our present industrial age. In the year 47 B.C. Julius Caesar reformed the old Roman calendar, which was based loosely upon the moon and sun, and based the new calendar entirely upon the sun. The year was fixed at 365¼ days and was divided into 12 months. The first month, Januarius, had 31 days, Februarius had 30, and the remaining months alternated between 31 and 30 days in that order. Since this would provide a year having 366 days, Februarius was reduced to 29 except on leap year when one day was added. This calendar was adopted in 46 B.C. and the Roman senate honored Julius Caesar by changing the name of one month from Quintilis to Julius. When Augustus Caesar rose to power in 44 B.C., he had the senate name a month after him so that Sextilis was changed to Augustus. In order not to be outdone by Julius, he had one day taken from Februarius and added to Augustus so that both Caesars would have 31-day months named after them. Complaints about the unequal lengths of the quarters led Augustus to make further changes. A day was taken from September and added to October, and one from November added to December.

In spite of all this, the year of the Caesars was about 12½ minutes too long. Over the centuries this discrepancy amounted to a considerable number of days. By the year 1582, an error of about ten days had accumulated from the time agreed upon for fixing Easter. Pope Gregory XIII therefore decreed that ten days should be deleted from the calendar; the day following October 4, 1582 should be called October 15, 1582. In order to prevent a recurrence of this situation, he further decreed that each centurial year divisible by 400 should become a leap year. This reduces the error to such a small amount that it would require 4,000 years to produce an error of one day.

A more logical solution to the calendar problem is advanced by Cable, Getschell and Kadesch in their book, *Science in a Changing World*. It is proposed that the calendar have 364 days instead of 365. Such a year would have an exact number of weeks and any given date would fall on the same day of the week year after year. The year would start on a Sunday and end on a Saturday. In order to provide for the 365th day required by our solar year, a Year-End Day would be added. This day would come between December 30 and January 1, and would not be a day of any week. It could not be referred to as Tuesday or Friday or any other weekday. It would simply by Year-End Day. Leap year would be taken care of by adding Leap-Year Day between June 30 and July 1. The two days would be kept out of the weeks and would become national holidays. Such a year would provide quarter- and half-years of equal length; each quarter would start on Sunday and end on Saturday; there would be 13 Sundays and 13 Saturdays in each quarter; the year would be perpetual since the days and weeks would stay in the same place each year. Such a calendar would be a welcome relief to commerce and industry as well as in our personal affairs.

How do we judge distances?

Ordinarily we use two eyes to see with and each views an object from a slightly different angle. We owe our ability to judge distance and depth to this fact. To illustrate this point, hold a pencil in each hand and with one eye closed, try to touch the points together. Put your hands behind your back between tries in order to reduce the effect of memory on the test. I am sure that you will agree that two eyes work better than one.

The depth perception that we have become accustomed to is the result of seeing partly around an object because of the separation of our eyes. With one eye covered, we see no better than an ordinary camera, reproducing an image that is flat and lifeless.

Why do ears "pop" during an airplane flight?

Almost everyone has at one time or another noticed pressure on the ear following a sudden change in altitude. This feeling is usually accompanied by a momentary reduction in the sharpness of hearing. A study of the cavity of the middle ear throws some light on the reason for this discomfort. This cavity is sealed at one end by the eardrum and is open to the atmosphere only through a relatively long narrow canal called the Eustachian tube. This canal is a very tiny one, and its walls are normally pressed closed, trapping a certain amount of air within the confines of the middle ear cavity. What happens, then, when we change altitude abruptly? As we go up, we enter a region of lower atmospheric pressure. As a result, the pressure within the middle ear is greater, for a time, than the pressure on the outside. This causes the eardrum to bulge out somewhat resulting in impaired hearing, a stuffy sensation and, in some cases, pain. This pressure difference becomes equalized, eventually, by the escape of air down the Eustachian tube. This escape is sometimes accompanied by a popping sound as the eardrum suddenly contracts to its normal position. Chewing, swallowing and yawning sometimes hasten this adjustment. When a plane descends to a lower altitude, the same symptoms are noticed and the explanation is completely analogous to that given above.

Why does a pressure cooker increase the speed of cooking?

Everyone is familiar with the fact that water evaporates. This means that water gradually changes from the liquid to the vapor state. At the surface of water, its molecules are constantly jumping from the liquid into the air above. If a pan of water is heated, its rate of evaporation increases. This is because heated water molecules have more kinetic energy (energy of motion) and consequently are better equipped to jump free of the surface. If the temperature of water is raised to 100° C. (212° F.) it begins to boil. If we try to heat water above 100° C., we find that the boiling becomes more violent and more vapor is produced but the temperature fails to rise. In boiling foods, therefore, 100° C. is the highest temperature available to us. If we want to boil foods at a faster rate, we must find a way to increase the temperature at which the water boils. This is exactly what the pressure cooker does.

The temperature at which water boils depends upon the pressure existing above it. Increase this pressure and water boils at a higher temperature. Reduce it and the boiling point of water is also reduced. In a pressure cooker, water and food are place in the cooker and the lid clamped on tight. As steam generates in the cooker it builds up the pressure above the water thereby allowing the water temperature to exceed 100° C. The pressure within the cooker increases steadily until it is high enough to operate the blow-off valve. At this point, any excess steam is allowed to escape. No matter how much heat is applied to the cooker, the steam pressure cannot exceed this amount. The temperature of the water in the cooker is fixed, therefore, at a point corresponding to the pressure so determined. Since this predetermined temperature is higher than 100° C., foods cook faster in a pressure cooker than in ordinary boiling.

At the top of Mount Everest, the highest mountain in the world, the pressure exerted by the atmosphere is only one-third of that at sea level. At this reduced pressure, water boils at about 72° C. As you can see, cooking by boiling can be a real problem at such elevations. The pressure cooker solves this problem by providing its own atmosphere and thereby determining for itself the rate at which foods can be cooked.

Does water dowsing really work?
Use of the divining rod dates back to antiquity. The Romans, for example, used a forked stick of willow called the *vigula divina* to aid in the location of hidden objects, and a similar gadget was in vogue during the sixteenth century to locate mineral deposits and

subsurface water. The relatively modern divining rod or "dowser" appears to have been first mentioned by Basel Valentine, an alchemist of the late fifteenth century. Despite its supposed magical power, it had begun to lose favor by 1650 and its use was turned to the detection of criminals and heretics. In addition to being more profitable, this use was less liable to exposure as a fraud.

The existence of modern electronic equipment for the location of oil and mineral deposits depends upon well-founded physical laws and, of course, these modern "dowsers" should not be confused with their predecessors of a superstitious age. Geiger counters, for example, are of great value in the location of uranium ore. As for water dowsing, the question is best left for the welldigger to answer. From a scientific point of view, the use of forked willow sticks should be confined to the barbecue pit.

How cold does it get at the bottom of a frozen lake?

Strange as it may seem, water remains at 4° C. (about 39° F.) at the bottom of most deep lakes during even the coldest part of the winter. Since ice may exist at the surface, this means that water at the bottom of the lake is warmer than water at the surface. This is just the opposite of conditions in everyday life where hot gases seem to rise. The fact that hot gases rise enables a chimney to carry carbon dioxide, normally heavier than air, up its tube and out of the house. Why then do the warmer particles of water fall to the bottom of the lake?

Most substances, including air, expand when heated and contract when cooled. This is why warm air rises to the ceiling of a room rather than falling to the floor. Water, on the other hand, is quite unconventional in this respect. Like most other substances, it contracts upon cooling until the temperature reaches 4° C., but at this temperature a reversal takes place. As the temperature is reduced below 4° C., water again begins to expand! By the time the temperature reaches 0° C., water is considerably lighter than it was at 4° C. At 4° C. water reaches its maximum density and therefore settles to the bottom of the lake. This odd behavior of water has extremely important effects in nature, chief of which is this ability of ice to form at the surface of lakes and rivers. If this were not so, our lakes would freeze at the bottom and probably remain frozen through the summer, thereby killing all marine life. In addition, water tends to collect in minute cracks of rocks, creating sufficient pressure upon freezing to split these cracks wide-open. This helps in the process of weathering, which is responsible for the formation of sand particles and, eventually, of our soil.

How long does a full-length mirror have to be?

Most full-length mirrors are about five feet long or nearly twice as long as they need be. Unbelievable? Well, let's examine the condi-

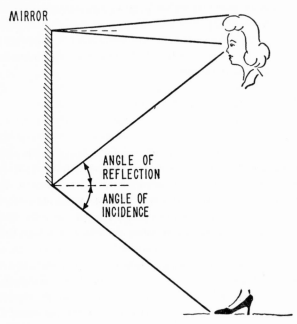

FIG. 2. A FULL-LENGTH MIRROR

Most full-length mirrors are about twice as long as they need to be. This young lady can see all of herself in a mirror that is about half as long as she is.

tions necessary for you to see your full length in a flat, vertical mirror. Let us assume that you are about six feet tall. Obviously, in order to see your face, you must place the top of the mirror almost as high as the top of your head. A point about halfway between your eyes and the top of your head should be just about right. In this way, light rays coming from the top of your head will be reflected back into your eyes. But what about your feet? Since the angle of incidence must equal the angle of reflection, light rays coming from your foot must meet the mirror about three feet (just about half your height) from the floor in order to be reflected into your eyes. It seems, therefore, that your full-length mirror must be about half your height in length, although a more parsimonious individual might reduce this figure by one-half of the distance between his eyes and the top of his head. If you are wondering whether the distance between you and the mirror has anything to do with this matter, you can test it for yourself with any full-length mirror. Cover the unnecessary portions of the mirror with paper and Scotch tape and then try viewing yourself from several different distances. You will see that it does not matter in the least how near the mirror you stand nor how far away you go.

What is trichinosis?

The trichina is a tiny roundworm that can live in the flesh of a number of animals. It passes from one to another when one animal devours another. In the case of man, the most important source of infection is pork. There may be as many as 85,000 of these worms in one ounce of ham. As in the case of beef tapeworms, they can be killed by cooking, but if eaten in raw pork, they cause painful illness. Condemnation of infected meat, as in the case of the beef tapeworm, is theoretically possible, but the worms are so minute in size that they must be searched for with a microscope. In the United States this procedure is not favored on the grounds that it is unwise to give official sanction to an examination that is not completely reliable.

Trichina worms do not reproduce in the human body. Thus, a person eating a large piece of poorly cooked, infected pork will be more seriously ill than one who eats a small piece. The ingested worms burrow through the walls of the intestine and find their way to the muscle tissue. Each worm that reaches a muscle becomes encapsulated and remains there for the rest of the life of the host. Symptoms of the disease may resemble "intestinal flu" in the early stages and muscular pains in the latter.

It is estimated that about 15 per cent of the population of this country is infected with trichina. Recently, it has been learned that exposure of pork to extreme cold kills the trichina. With continued cold treatment, the incidence of this infection should be reduced in the future.

How does a television set pick out the "right" picture?

Your radio or television receiver must distinguish between hundreds of different stations transmitting at the same time if it is to operate properly. This selection is accomplished in the *tuner* by a process called *resonance*.

If we place a number of different-sized drinking glasses on a table and strike each with a spoon, we will find that each gives off a different tone or pitch. This is because each vibrates at a different rate. The rate or frequency at which an object will vibrate is called its natural frequency of resonance.

To get a clearer idea of resonance and frequency, suspend a small weight from a string about a yard long and hang the string from a nail so that the weight is free to swing back and forth. You now have a simple pendulum which you can start swinging by means of a gentle tap. You will notice that the pendulum takes exactly the same length of time to make each swing even though these swings become progressively shorter in length. The number of complete swings that take place in a given period of time is called the natural frequency of the pendulum. Now that the pendulum is in motion, give it a very light tap each time it reaches the end of its swing. You will soon have it swinging violently back and forth. Try tapping it at some point other than the end of its swing and you will cause the swing to slow down. In the first case, each tap added a small amount of energy to the swing which caused it to increase to a violent excursion. In the second case, the tapping merely impeded this swing. Therefore, to get the most efficient transfer of energy from your hand to the pendulum, the tapping must occur at the natural frequency of resonance of the pendulum. When this condition exists, the tapping is said to be in resonance or in tune with the swing of the pendulum.

If two bodies have the same natural frequency and one of them is set into vibration, a small amount of energy will be transferred to the other body, causing it to vibrate at the same rate. This effect can be illustrated with the help of two similar milk or pop bottles. Ask a friend to hold one bottle with the opening close to his ear without obstructing the opening. Now blow strongly across the opening of your bottle to produce a loud, clear tone. Part of the energy from this note will reach the second bottle, causing it to vibrate at the same rate or frequency. Athough the note is weaker, your friend should have no trouble in hearing it distinctly.

It is a similar effect which enables a television or radio receiver to select a desired station from among the numerous ones cluttering up the air waves. When you rotate the tuning knob, you are actually changing the natural frequency of the radio set to agree with that of the desired station. Were it not for this ability to change the receiver's resonance, you would see all of the pictures at once on your television screen.

Who invented the wheel?

The inventor of the wheel will probably never receive recognition for what was one of the greatest inventions of all time. Like the bow and arrow, it has come down to us from prehistoric times as a monument to man's ingenuity. It is interesting to conjecture upon the steps that might have led to this discovery. Imagine a prehistoric man dragging his game through the forest on his way home. Along the route he notices the trunk of a small tree across his path. At this point he wonders, "Around or over?" Trusting to luck, he decides to drag his game over the log. Imagine his surprise to find that the log rolls under the animal and lightens his load tremendously. He reasons that the log can be cut into a short length and used over and over again until he gets back to the cave. This method may have been used for thousands of years until a prehistoric Galileo or Edison began to analyze the problem. A hollow log, an axle, and there you are—a wheel!

As we reflect upon the countless centuries required to produce these early inventions we must bear in mind that men of that era lacked our most important heritage, the ability to read and write. It is this heritage which makes it possible for succeeding generations to stand upon the shoulders of their predecessors, adding their bit to the total store of knowledge, rather than rediscovering that which was already known in the past.

What are enzymes?

Although scientists do not agree on what life *is,* they do agree that, whatever it is, it exists or occurs inside of cells. It is within these cells that living processes take place. There are literally hundreds of these processes upon which life depends, and the sum total of them all is called metabolism. Many of these metabolic processes, or chemical reactions, have been studied, with the interesting discovery that almost every key reaction is brought about with the help of an *enzyme*. Furthermore, enzymes are not used up in these processes, but merely serve as catalysts, or helpers, without which the chemical reaction would take place much more slowly, or not at all.

All enzyme molecules are protein molecules. To date, scientists have not succeeded in synthesizing a single protein molecule in the laboratory. Nevertheless, a great deal has been learned about the properties of enzymes by extracting them from cells, purifying them, and studying their behavior in test tubes. In this way, the effect of a given enzyme can be determined by noting its effect upon various organic compounds which are normally found in the body. From studies of this type, it has been determined that enzymes have the following characteristics:

1. Enzymes are destroyed by moderately high temperatures. Elevated temperatures convert them to other compounds and destroy their effectiveness.

2. Enzymes are sensitive to the acidity of their surroundings. Some work best in acid solutions, some in alkaline solutions, while others require neutral solutions.

3. Enzymes are very particular about the type of chemical reaction that they will help along.

While science has learned a great deal about enzymes and their effect on life-giving chemical reactions, their true nature and their mechanism of operation still remain a mystery.

Why is it difficult to dissolve sugar in iced tea?

A solid going into solution dissolves only at its surface. In this respect, the process is similar to evaporation of liquids. As sugar dissolves, a layer of the highly saturated solution tends to blanket each sugar crystal, preventing underlying layers from dissolving. This blanket cannot be removed easily by stirring because this extremely thin layer moves along with the sugar crystal even under agitated conditions. As a matter of fact, it has been proved by scientists that the layer of a moving liquid which is in contact with a fixed surface, such as the inside of a water pipe, does not move at all. Then what does enable the sugar to dissolve? The answer lies in the fact that all molecules are in motion. At high temperatures, this molecular motion is increased because of the higher energy that each molecule possesses. This increases the rate at which the dissolved sugar particles move away from the sugar crystals, thereby enabling additional unsaturated tea to come in contact with the sugar. At lower temperatures, the converse is true and the sugar goes into solution at a lower rate.

Why does ordinary table salt refuse to pour in humid weather?
Some substances, when left exposed to air, soon absorb enough water to become wet, and dissolve themselves in this water. Such substances as calcium chloride, a white solid, are often used on dry dirt roads and tennis courts in order to absorb moisture and keep the dust down. Sodium hydroxide (known commonly as lye) and magnesium chloride are other examples of water-absorbing substances.

Many materials, such as wool, silk and tobacco, absorb moisture, but to a much lesser extent. Pure table salt, sodium chloride, is unlike all of these substances in that it does not absorb water to any extent. In the pure state, it pours just as easily in damp weather as in dry. In its normal form, however, table salt contains small amounts of magnesium chloride and calcium chloride. These substances absorb great quantities of water from the air and cause table salt to "cake" during humid weather.

What are shooting stars?
About ten billion meteors, or shooting stars, enter the earth's atmosphere each day. Fortunately for us, their average size is so small that you could hold thousands of them in the palm of your hand at one time. Actually, they are no more than specks of celestial "junk" that happen to enter the earth's atmosphere. Their speeds vary from about forty-five miles per second to eight miles per second depending upon whether they meet the earth head on or must catch up with it before entering its atmosphere.

Meteors, of course, aren't shooting stars at all. Their brilliance is due to the heat generated in their rapid flight through our atmosphere. They light up to incandescence for a brief second, only to disintegrate just as suddenly into dust.

Even though most meteors are very small, the average particle gives off the same amount of light as about 100,000 ordinary light bulbs. The total weight of these ten billion meteors is not much more than twenty pounds, so the earth's weight is increased by them at the rate of about a pound an hour.

Several swarms of meteors travel around the sun in orbits, just as the earth and other planets do. When one of these swarms and the earth come together in space, a meteoric shower is the result. Many of these have been quite spectacular. Some of the recurring showers appear as follows: August 11, April 20, a period from November 20 to 27, a period in early May, October 19 and December 11. If you look for shooting stars on some of these nights, you might be rewarded by seeing a whole cloudburst of them.

Why do the tides rise and fall?

The connection between the moon and the level of the ocean has been recognized for at least two thousand years. It was not until the discovery of the law of gravitation in 1687, however, that a scientific explanation of the tides was proposed. Newton showed mathematically, at that time, that the attractive force of the moon and, to a lesser extent, of the sun would explain the existence of the tides.

The moon attracts each particle of water on the earth. Since the earth is round, however, the distance between the moon and the earth varies from point to point on the earth's surface. The pull of the moon must vary, therefore, in this same manner. This causes the water directly under the moon to heap up somewhat, producing a high tide at that point. It is relatively easy to understand the existence of this high tide, but at the same time there is also a high tide on the opposite side of the earth. Let's see how Newton's law of gravitation helps to explain this situation.

The law states, in effect, that the water closest to the moon will receive the greatest pull, the water furthest from the moon will receive the lightest pull, while the water midway between will receive a pull somewhere between these two in strength. Since the water on the side of the earth opposite the moon receives the lightest pull, it will also produce a high tide.

As the earth rotates about its axis, the point of highest water stays almost directly under the moon, causing two high tides and two low tides each day. As a result, there are two tidal waves moving continually around the earth. If the moon were stationary, the high tides would occur exactly twelve hours apart; but since the moon is moving, the tides are about twelve hours and twenty-five minutes apart.

In addition to the moon, the sun also makes a contribution to the production of the tides. Its effect is about five-elevenths of that of the moon. It tends to produce high tides at noon and midnight each day, with low tides at six o'clock in the morning and evening. Since the solar effect is the same each day, while the lunar tides occur about fifty minutes later each day, it is evident that they sometimes reinforce each other and sometimes subtract from each other. At periods of reinforcement, during the time of the full moon and new moon, we have *spring tides*. A week later, during the first and third quarters of the moon, the two effects are in opposition, and we have a *neap tide*.

What causes lightning?

There are many theories concerning the nature of lightning, and although there is agreement on the main points, the details involved are still open to debate. The general idea seems to be that during a thunderstorm, rain, snow or hail becomes charged with electricity in falling through the swift upcurrents of air that usually accompany such storms. This concept was originated by Michael Faraday, the great English scientist. He showed that a great electric charge could be generated by a steam-driven spray of water directed against a water surface. If this idea is applied to the thunderstorm, the falling raindrops will develop a charge and, in doing so, will cause the upward currents of air to develop an equal but opposite charge. The raindrops drop to earth, but the electrically charged air rises to the top of the clouds. This causes the clouds to develop a great electric charge and the stage is set for an eruption. As soon as the charge becomes great enough, the insulation of the air breaks down and a flash of electricity leaps from one cloud to another, or from cloud to ground. It is this flash of electricity that we call lightning.

How high can a balloon rise?

A flying balloon is essentially a light, strong, airtight bag filled with a gas that is lighter than air. It will rise if its weight when added to the weight of the gas it contains is less than the weight of the air it displaces. Helium, which is usually used for the purpose, weighs only about one-seventh as much as air. This provides the lifting power that is required. Since air is very compressible, the lower layers are compressed by those existing above them. As a result, air has its greatest density at sea level. As we go higher and higher, air gets thinner and thinner until there is none left. As a result of this progressive change in density, the lifting power of the balloon gets lower and lower until the weight of the balloon plus helium is just equal to the weight of air displaced. When that altitude is reached, the balloon stops rising and floats as though it were on the surface of a liquid.

Why does the pitch of a railroad train whistle appear to diminish as the train rolls past?
A person standing near a railroad track hears a distinct lowering of pitch in the whistle of a train as it rushes past him. If two cars pass each other on the highway while the horn of one is sounding, the occupants of the other notice a change in pitch as the two cars pass each other. These and many other similar observations illustrate a principle applicable to all kinds of wave motion. It was developed by Christian Doppler and is known as the *Doppler effect*. To help understand this principle, imagine a number of water waves traveling along the surface of a lake. To a person on the shore, these waves seem to travel at a uniform rate in a given direction. To a person in a canoe, however, they appear to be moving at a different rate, either faster or slower, depending upon the direction of motion of the canoe. If it is moving against the waves, more waves than normal will pass the canoe in one second. If it is moving with the waves, the velocity of the waves will seem to be reduced with respect to the canoe.

This same effect can be applied to sound. Suppose we are moving toward a sound source, such as a ringing bell. A bell has a given pitch because it emits a certain number of waves per second. Our ear "counts" the rate at which such waves *reach the ear* and automatically ascribes a certain pitch to the sound. If we move toward the sound source, our ear receives more than a normal number of sound waves per second and the bell seems to have a pitch which is higher than normal. If we move away from the bell, our ear receives less than a normal number of sound waves and the bell sounds lower in pitch than normal. As a matter of fact, if we should move away from the bell at the speed of sound, we would hear nothing since the sound waves could never quite catch up with us.

What causes motion sickness?

Motion sickness results from the fact that the semicircular canals of the inner ear respond to body motion. If both canals are destroyed, an individual becomes deaf and his sense of equilibrium is seriously impaired. In a darkened room, without the aid of eyesight, he cannot move around with any degree of safety.

The sense of equilibrium is not an unmixed blessing, however, since the perception of certain kinds of motion can lead to motion sickness. It is probably that other factors, such as vision, memory of past experiences and general physical condition, may contribute to the effect. In any event, practically all human adults are subject to motion sickness to some extent.

Why is the sky blue?

Each of us has had an opportunity to view the beautiful blue of the midday sky and the striking red and yellow hues that sometimes occur at sunset and sunrise. Have you ever wondered where these colors come from? To understand these phenomena, we must recall that sunlight is a mixture of light of all colors and that such a mixture appears white to the human eye. As this light passes through our atmosphere, it is partially scattered by molecules of air, as well as by dust, water vapor and other impurities present in air. If light of all colors were scattered to the same extent, the sky would appear white, as would our sunsets. Since the sky is blue, and not white, there must be a physical explanation to account for it.

Scientists have established that the relatively short waves of violet and blue light are scattered about ten times more than the longer light waves corresponding to red. It is this selective scattering which accounts for the blue color of the sky. The long wave-length red rays tend to go straight through our atmosphere while the short wave-length blue rays are scattered from their original direction by the air, water and dust particles overhead. It is this scattered light that we see when we look up into the sky. Ten or twenty miles straight up in the stratosphere, there is very little air left overhead and practically no light is scattered downward. An observer at that altitude sees only the empty black void of space, broken only by the light of stars.

What are atoms?
An atom is the smallest particle of an element having the chemical properties of that element. Atoms are, for the most part, empty space. There are ninety-two different kinds of atoms found in

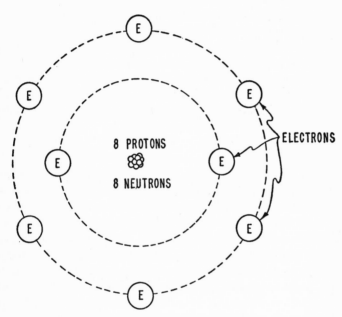

FIG. 3.   AN ATOM OF OXYGEN
This atom consists of a nucleus of 8 protons and 8 neutrons, around which circulate 8 orbit electrons.

nature, one for each of the ninety-two chemical elements. In order to better understand its inner workings, let us take an imaginary trip from the outside toward the center of an atom. As we get close to it we notice a number of electrons revolving rapidly in various orbits about the center. Upon closer examination we notice that all of these electrons are exactly alike and all are charged negatively. Furthermore, if we add up the weights of all of these electrons, we find the total to be very small in comparison to the weight of the atom. If we count the number of electrons, we find that the number depends upon the kind of atom we are investigating. If it happens to be hydrogen, there will be only one; if it is uranium (which has the greatest number of whirling electrons) there will be ninety-two; if it is copper, there will be twenty-nine. Still, no matter what kind of atom we study, the total weight of the electrons will be only a few hundredths of a per cent of the total atomic weight.

At this point, some questions are probably popping up in connection with our atom. Where, we might wonder, is the remaining mass of the atom? Why is it that these negatively charged electrons, which must undoubtedly repel one another, don't fly off into space? What holds the atom together?

The center, or nucleus, of the atom holds the answer to all of these questions. The nucleus is a tiny particle, no larger than an electron, but so heavy that it contains almost all of the weight of the atom. It carries a positive charge exactly equal to the total negative charge of the electrons rotating about it. It is the attraction of this positive charge that restrains the rapidly whirling electrons from leaving the atom. The atom as a whole is electrically neutral because it contains as much positive charge (on its nucleus) as it does negative (on its electrons).

Although an atom is extremely small, the electrons and the nucleus are so much smaller still that there are tremendous empty spaces between particles. If the atom were increased in size millions of times, so that the nucleus were the size of a baseball, the trip from the outermost electron to the center would be about half a mile long.

Now that we have a general picture of the structure of the atom, let's investigate the nucleus. Just what is inside the nucleus? According to accepted theory, it takes two kinds of particles to make up any nucleus, from the simplest to the most complex. These particles are called protons and neutrons. A proton has a weight about eighteen hundred times that of an electron, and a positive charge equal to the negative charge on an electron. A neutron has a mass about equal to that of a proton, but has no charge at all. We usually think of a nucleus as consisting of closely packed protons and neutrons. Change the number of either particle and you have a different kind of nucleus. Although there are only 92 different chemical elements found in nature, there are about 250 kinds of nuclei. Evidently some elements must have more than one kind of nucleus. Such elements are called isotopes.

Of course, this simple picture of a nucleus gives rise to a lot of unanswerable questions. How can so many positively charged protons stick together in such a tiny confined space? Are protons and neutrons motionless, or do they move around in the nucleus? The number of protons is equal to the number of orbit electrons, but what determines the number of neutrons?

These questions are indeed embarrassing, and as a matter of fact, represent a few of the major problems before nuclear physicists today. Fortunately, much has been learned about the behavior of nuclei in spite of these puzzling questions. For our purposes, the concept of a nucleus as a close-packed cluster of protons and neutrons will be adequate. A further description of nuclei will be undertaken under the discussion of isotopes.

Why does baking powder make cakes rise?

As every housewife knows, the difference between a light, fluffy cake and a discouraging failure is often merely forgetting to add one or two teaspoons of baking powder. When added in the correct amount, baking powder produces many tiny bubbles in the cake which cause it to rise to the desired degree. Baking powder accomplishes this feat by producing carbon dioxide gas under the influence of heat. The gas bubbles so produced are able to remain within the batter if the temperature of the oven and the consistency of the batter are correct.

All baking powders contain ordinary bicarbonate of soda. In addition, they contain an acid-forming substance such as cream of tartar, alum or calcium acid phosphate. When bicarbonate of soda and one of these acid substances are dissolved and heated, carbon dioxide gas is liberated as discussed above. The acid substances differ from each other in the rate at which carbon dioxide is liberated at different temperatures. Some so-called double-acting baking powders contain two of these acids, one of which reacts at a lower temperature than the other. This increases the chances of a favorable outcome and helps to maintain the good disposition of the housewife.

In freezing ice cream, why do we add salt to ice?

Anyone who has made ice cream at home knows that a mixture of ice and salt will produce a temperature low enough to freeze ice cream while the use of ice alone would not be adequate. This is because salt, on going into solution, absorbs heat from the ice, thereby effectively reducing the temperature of the solution. The reason for this absorption of heat can best be explained by an example.

If we place a crystal of hydrated cupric sulfate in water, it slowly becomes smaller and finally disappears, while the liquid in the immediate vicinity turns blue. The dissolved particles leave the surface of the crystals and move about in the water as though they themselves were a liquid. They eventually diffuse throughout the water and appear to forget that they are, in reality, particles of a solid. This increased mobility, or freedom of action, is due to the fact that the dissolved particles have absorbed heat from the water, causing a reduction in temperature of the solution. The same effect takes places when salt is mixed with ice, and it is this process which produces the low temperature we observe.

A product on the market uses this principle to chill beverages on picnics when ice may not be available. All that is required is to place the bottle to be cooled in a container filled with water and a mysterious powder. As soon as the powder (photographer's hypo) is dissolved, the water becomes ice-cold and chills the bottle. This method may be expensive, but it's a lifesaver in an emergency.

How do we taste food?

Our sense of taste depends upon organs called taste buds which are located chiefly on the tongue. Each taste bud consists of between ten and sixteen taste cells which are connected to the nervous system. All of our taste buds can be divided into four groups, those which respond to salt, sweet, bitter and sour substances. The taste ascribed to a given food depends upon the relative response of each of the four types of taste buds.

It is interesting to observe that the many preferences and prejudices concerning the taste of foods have their basis in only four basic taste sensations. It would seem that psychological factors are very important in determining our preferences and prejudices. Getting back to the biological aspect of taste, we find that the tongue is mapped out into more or less distinct taste areas. A sweet-tasting area is located at the tip of the tongue; a bitter-tasting area at the back; and sour-tasting areas along the sides. The salt-tasting buds are located evenly over the entire surface of the tongue. These are the only flavors which we are capable of tasting. The much-maligned onion actually has no taste at all. We smell onions rather than taste them. The same is true of many foods which seem to have no taste when the nose is obstructed during a cold. It is for this reason that the senses of taste and smell are often called the partner senses. They work together efficiently and without our conscious knowledge to produce our reactions to foods.

What is radar?
During World War II, even the word "radar" was secret. Strange as it may seem, the word itself was so highly classified that it could not be used by scientists except as required in their work. Then, for a while, the restrictions were lifted but not even the vaguest hint as to its meaning could be disclosed. It was during this period that an unknown humorist circulated the information that "radar" is "radar" spelled backward.

We know today, however, that radar is used to detect and indicate the location of aircraft, ocean-going vessels, and targets for our bombers. It is used to direct the fire of antiaircraft guns, to help night fighters find their targets in the dark, and to guide planes to a safe landing when their pilots cannot see the ground. To do these things, a radar set must have three characteristics: it must be able to *detect* an object at a great distance; it must be able to determine the *direction* of the object from its own location; and it must be able to measure the distance or *range* of the object. Thus, a radar set uses radio waves to detect a distant object, to measure its direction, and to determine its range.

Now let us take these items up one at a time. First, let's discuss how detection is accomplished. Have you ever shouted toward a cliff and waited for the echo to return? Some of the sound waves strike the distant object and bounce back toward their source. This is a perfect illustration of how a radar set works. In radar, radio waves are echoed back instead of sound waves. Therefore, to detect an airplane flying one hundred miles away, we must send out a burst of radio waves and then wait until an echo returns. An operator sitting before a radar screen will see many such echoes from hills, buildings and even tall trees. How, then, can he detect the airplane? In the case of a fixed object, such as a building, each succeeding echo ends up at the same point on the screen. The screen, therefore, is actually a radar map of the surrounding terrain. When an airplane flies through this terrain, the echo, or "pip," is seen to move with respect to the fixed terrain. It can be detected, therefore, by its motion.

To determine the direction of an object, the radar set merely sends out bursts of radio waves in one direction at a time. As a matter of fact, the radar set's directional antenna rotates continuously. This enables the set to sample segments of the search area one at a time. As soon as a moving object is detected, the antenna can be stopped "on target" and it will then point directly at the moving object.

To determine the range of the target, engineers make use of the fact that radio waves take a definite length of time to reach the target and bounce back to their origin. The farther away the target happens to be, the longer the length of time required for the round trip. A radar set measures this time very accurately and converts the time lapse to feet or miles for presentation to the operator. Distances can be measured by modern radar sets with an accuracy of better than ten feet.

By combining detection, direction and range, a radar set provides all the information we need to know to determine the location of moving objects.

Are blue eyes really blue?

In general, the color of our eyes is due to pigment on the front area of the iris. The iris is the portion of the eye containing the pupil. It is located in front of the lens but behind the transparent cornea. Brown-eyed people have brown coloring matter on the front surface of the iris. In the case of blue eyes, this dark coloring matter is located on the *rear* surface of the iris. The blue effect is actually an optical illusion. It only *appears* to be blue. In reality it is reflected from the rear inside surface of the eyeball. When the light reflects from this surface, the effect is blue.

Albino eyes are caused by the complete absence of pigment from either the front or rear surface of the iris. The albino effect is produced by the reflection of light from blood vessels in the eye.

Is margarine as wholesome as butter?
Margarine is an inexpensive and wholesome table and cooking fat. Although many Americans seem to prefer butter, it is nevertheless true that margarine is just as digestible. As far as food value is concerned, margarine that is fortified with vitamins A and D is the equal of butter. Most margarines on the market contain added amounts of these vitamins.

Margarine is one of a number of substances called hydrogenated oils. When liquid oils are heated in the presence of hydrogen and nickel, hydrogen unites with the oils to produce a solid fat. The nickel helps the chemical reaction to take place without actually entering into the reaction. In hydrogenating such oils, it is common practice to stop the process before all of the oil molecules have reacted with the hydrogen. In this way the hardness of the product can be regulated to the desired degree.

Many of the common fats used in cooking and baking are hydrogenated oils. Margarine usually contains hydrogenated soybean, coconut, cottonseed or peanut oil. It is colored with a harmless yellow coloring matter and is usually churned in skim milk to give it a butter taste. Sometimes beef or lard oil, called oleo, is added.

The reason behind all of this hydrogenation is the American housewife, who seems to prefer solid fats to liquid oils for cooking.

Why does steam disappear before our eyes?
We all know that water evaporates even though we have never seen water vapor leave its surface. Similarly, ice and snow will slowly evaporate without melting if the weather allows them to remain on the ground long enough. Since water vapor is invisible, we do not see this evaporation take place. Steam is merely another name for water vapor and therefore it must be invisible. Yet we speak of seeing steam coming from a locomotive whistle or from the spout of a teakettle. What we see is not the steam itself but rather tiny droplets of water which are formed when the steam cools sufficiently to condense. You can convince yourself of this fact by observing a teakettle of boiling water. Immediately in front of the

spout you will see no indication of steam at all but an inch or so away you will see a cloud of condensed vapor. Steam, or water vapor, consists of molecules which are widely separated and invisible to the eye. When they have condensed into droplets, the droplets are large enough to be seen. These droplets then disappear because of rapid evaporation from their relatively large surface areas.

Why is carbon monoxide so poisonous?

Carbon monoxide is a gaseous compound consisting of one atom of carbon for each atom of oygen. It is a cumulative poison in that it collects slowly in the body until a dangerous concentration is reached. The physiological properties of carbon monoxide are such that it combines with the hemoglobin in the blood to form a very stable compound. This renders the hemoglobin useless as a carrier of oxygen throughout the body. The formation of this compound takes place gradually—slowly building up its concentration in the blood. When the concentration reaches about 40 per cent, the victim collapses and a further increase may mean death. One part of this odorless gas in ten thousand parts of air is usually sufficient to produce nausea and headache, while one in eight hundred can kill in less than half an hour.

Air does not normally contain carbon monoxide but there are many ways in which it can contaminate the air we breathe. The exhaust fumes from internal combustion engines contains from 1 to 10 per cent carbon monoxide. Improperly banked coal fires are rich in carbon monoxide and present a particularly dangerous situation. This gas should not be confused with its harmless relative, carbon dioxide, which is the gas present in soda pop. Although they consist of the same elements, they are as different as night and day. Carbon dioxide actually stimulates breathing and is sometimes given to patients suffering from carbon monoxide poisoning in order to produce this result.

What are isotopes?

You will recall that the nucleus or center portion of an atom consists of protons and neutrons. The number of protons in a given atom is always equal to the number of electrons whirling about the nucleus. The number of neutrons, on the other hand, appears to vary, even among atoms of the same element, such as copper. A very general rule about nuclei states that a stable nucleus, that is, one which will probably last indefinitely, must contain at least as many neutrons as it does protons, and possibly a few extra. Copper, for example, may contain either of two kinds of nuclei. Each has 29 protons, but one contains 34 neutrons while the other contains 36 neutrons. In either case the atom will have exactly 29 electrons whirling about the nucleus. The chemical properties of the atom depend only upon how many of these electrons are present; the nucleus has no effect whatsoever on chemical properties. So both kinds of atoms, one with 34 and the other with 36 neutrons, are atoms of the same element, copper. These two slightly different atoms are called the two *isotopes* of copper. They are chemical twins, differing only in the weight of the nucleus.

Scientists have learned how to make copper with other than the "right" number of neutrons in the nucleus. These are the so-called radioactive isotopes of copper. If we were to make some of these isotopes of copper, we would soon find that nature has some fairly definite rules about the number of neutrons that should be used with our 29 protons. If we try 35, the nucleus is unstable and will "last" only about 13 hours. If we try 33, the nucleus will last only 11 minutes. If we try 37, the nucleus will last only about 5 minutes.

Scientists do not know exactly why the number of neutrons has such a profound effect upon the stability of the nucleus. They have discovered, as mentioned above, that new unstable nuclei can be manufactured in the laboratory by bombarding atoms with high-speed neutrons. Hundreds of such nuclei have been produced thus far, but most of them are relatively short-lived. These radioactive isotopes are extremely useful, however, in medical investigations and treatments. They constitute a new and effective tool in the search for knowledge.

How do desert animals withstand the heat?
Desert animals have solved the heat problem by learning to dig burrows and live underground during the daytime. Most desert animals, such as the pocket rats and mice of our American desert, can't long survive when subjected to the desert's noonday sun. The highest temperature they seem able to tolerate is about 95° F.

They have been able to colonize our parched deserts only because of their burrows and nocturnal habits. The temperature of the air in a typical burrow never rises above about 90° F. and a more usual figure would be nearer 80° F., even on a hot day. Temperature differences as great as 55° F. have been measured between the temperature of the ground and that of the air in a burrow only a few inches below the ground. The same tempering effect takes place when the weather turns cold. During the night, the temperature within a burrow may be twenty-five or thirty degrees warmer than the surrounding air. In addition to the constancy of temperature within the burrow, the relative humidity tends to rise due to the accumulation of water vapor from the animal's body. His underground home thereby protects the animal from the effects of both high temperature and dehydration.

Did people believe that the world was flat at the time of Columbus? This is a popular misconception about Columbus and the people of his time. All of his intelligent contemporaries—especially mariners—knew that the world was round. His sailors didn't fear falling off the edge of the earth; they were afraid of running out of provisions on the return voyage. Columbus was well armed with scientific data. While marooned for a time on Jamaica, he sternly told the natives that they must give him food or he would turn off the sun! True to his word, on February 29, 1504 Columbus "produced" a solar eclipse which shocked the frightened natives into a more co-operative mood. Thus Columbus beat Mark Twain's Connecticut Yankee to the trick by several centuries.

People have known that the earth is round for thousands of years. By 200 B.C., Eratosthenes had succeeded in measuring the earth's diameter and circumference by scientific means and with considerable accuracy. He found by chance that on June 21, the longest day of the year, the noonday sun was directly overhead in Assuan, a town near the first cataract of the Nile. At noon on this day, the sun shone directly into a deep well and was mirrored back from the water's surface. With this knowledge, he made it a point to be in Alexandria, some 500 miles due north, on a succeeding June 21. He then measured the angle that the noonday sun made with the vertical at Alexandria. He found this angle to be 7½ degrees. This information, and a knowledge of geometry, enabled him to calculate the earth's circumference as 25,000 miles. His estimate of the earth's diameter was even more accurate, 7,850 miles—only about 50 miles too short.

How was coal formed?

Scientists believe that millions of years ago, during the Carboniferous period, the earth was covered by a luxuriant vegetation far more dense than that which is found today even in tropical jungles. During one of the many upheavals that occurred in this period, the earth sank in certain places producing swamps which were later completely submerged. As streams carried mud, sand and clay into the areas, the submerged vegetation was subjected to increased temperatures and pressures.

It is believed that partial decomposition of this vegetation changed it first into *peat.* Extensive peat bogs are found in Pennsylvania, Michigan, Wisconsin and many other states. These bogs sometimes reach a depth of several feet of partially decomposed mosses, sedges and other forms of vegetation. Although its fuel value is rather low and it burns with a smoky flame, peat may be dried and used as fuel. The peat bogs of Ireland and elsewhere represent coal in its early stages of formation.

The next step in the formation of coal is believed to be the changing of peat into *lignite,* which is sometimes called brown coal because of its color. It shows the structure of the wood from which it was formed and the branches or twigs which are sometimes found in lignite indicate its origin. It burns with a smoky flame and yields less heat than coal.

Further decomposition and pressure, in the absence of air, results in the formation of soft or *bituminous coal.* It has a higher percentage of carbon and a greater heat value than lignite and is mined extensively in many parts of the United States.

There is no doubt of the vegetable origin of hard coal or *anthracite.* It appears to have been subjected to still higher pressures and temperatures than the forms discussed previously. It has the highest percentage of carbon and the lowest percentage of volatile materials. Anthracite appears to be the last of a series of decompositions which started with wood. In this process, the heat value per pound of coal has approximately doubled that of the wood from which it was formed.

Do animal instincts really exist?

All of us are familiar with the marvelous things that insects do. We have heard of the complex social organization of the ants and of the complicated structures insects can build. Their actions are all the more remarkable when we consider that they are not the product of learning. A spider knows how to build the right kind of web for its species without ever having seen one. The larvae of the silk-worm moth spin cocoons of a single silken thread more than a thousand feet long with no prior experience. The female *Pronuba yuccasella* moth knows it should lay its eggs in yucca flowers without being told. Why do animals and insects behave in this manner?

There has always been a tendency, when animal behavior cannot be analyzed, to ascribe this behavior to "instinct." The term itself is not particularly dangerous, but its implications are. We must be continually on guard against taking refuge in a word rather than searching for the answer. "Instinct" is certainly not the answer to the mysteries of animal behavior.

An interesting example is given by one of the wasps at egg-laying time. The wasp catches and stings a spider, lays his paralyzed prey aside, digs a hole, lays an egg in it, places the spider in the hole, and then seals the hole. Here is a reasonably complicated activity, consisting of seven distinct acts, always performed in the same order. In order to learn more about this behavior, someone tried the experiment of removing the spider while the wasp was digging the hole. After finishing with its digging, the wasp laid an egg and came to get the spider. Since the spider was missing, he ran off, found and stung another spider, dug a new hole, laid a new egg, put the spider in the hole, and sealed the hole. The first egg was completely forgotten and left to die of starvation when it hatched. This is a key to the behavior pattern of the wasp. This behavior is not really one complicated activity but a series of simple activities or reflexes, each step of which stimulates the next. The first step is probably the secretion of a hormone by the reproductive organs. This provides the stimulus for the "catch the spider" response which, in turn, becomes the stimulus for the "sting the spider" responses and so on.

32

This is why the wasp must repeat the whole process from the beginning if the spider is stolen.

The actions of a different kind of wasp have been described by J. H. Fabre. This wasp has a habit of paralyzing a cricket and then storing it in a hole in the ground. When bringing a cricket toward a predug hole, it flies close to the hole, lands, and drops the cricket. Then it goes down into the hole for an inspection, comes out, picks up the cricket, and stuffs it into the hole. In an inquisitive mood, Fabre waited until the wasp was down making his inspection tour and then moved the cricket a short distance away. When he came out the wasp was somewhat disturbed by the theft but soon found the cricket and brought it back to the vicinity of the hole. Instead of putting the cricket in the hole, the wasp *went down for another tour of inspection*. While it was in the hole, Fabre again moved the cricket with the same reactions by the wasp. The weary Fabre repeated this experiment until he was quite tired of the affair but each time the wasp insisted on revisiting the hole that he must have known by heart. Fabre concluded that a series of inborn reflexes cannot easily be interrupted. This type of behavior is due to a series of inborn reflexes which always work more or less automatically. It is probable that many other so-called instincts such as the migration of birds can be explained in the same way.

How does an absent spider know that an insect has entered its web?
The female spider, who does all of the spinning, is equipped with tubelike appendages which eject a silken thread only about four-thousandths of an inch in diameter. Before she leaves her web, she spins a silken telegraph cable which is attached both to the web and to her own body. When a fly or beetle strikes the web, she feels the vibrations in her cable and hurriedly returns to take charge. She binds the intruder in her threads and injects a suit-able quantity of poison. When completely safe, she proceeds to suck away its blood.

Another interesting aspect of the spider's activities is the spinning of a web. The spider displays amazing skill in designing perfect geometric patterns in her work. She instinctively follows circles, triangles and circumferences in the construction and even takes the trouble to remove temporary spans that were used to expedite the work. The web must be architecturally and aesthetically sound before she considers her work complete.

Can a worm learn?

Learning can be defined as a change in a behavior pattern as a result of experience. The two methods by which animals (including worms) appear to learn are called trial and error and conditioned response. Of course, the degree of learning possible depends upon the complexity of the animal's nervous system. Man has the highest capacity for learning while some of the lower animals learn only with great difficulty. But in any event, many scientists agree that the fundamental processes are the same, no matter who does the learning. These processes are best explained by examples and several classical experiments are described below.

In 1900, Ivan Pavlov, the famous Russian physiologist, noticed that the sight of food could make a dog's mouth water. Pavlov knew that dogs, like most animals, have an inborn reflex in which saliva flows in response to food. He wanted to find out if this reflex could be changed, so he set up an experiment in which he placed a dog in a small bare room with a minimum of distracting elements. After the dog had settled down, a bell was rung and soon thereafter some food was lowered into the room. When the meal was finished, the plate was quietly removed. This procedure was repeated several times and as Pavlov watched through a small hole in the wall he began to notice a change in the dog's behavior. After a few experiences of having the bell ring just before the food arrived, the dog began to show excitement at the sound of the bell. It licked its lips and began to salivate just as though the food had arrived. Pavlov called this response a conditioned reflex, and his work in this field proved to be a substantial contribution to the study of animal learning.

Yerkes performed an experiment designed to investigate the trial-and-error type of learning. He constructed a small T-shaped passage just large enough for a worm to crawl through. He started the worm in at the bottom of the T. If the worm turned left upon reaching the arm of the T, it would have to pass over rough sandpaper beyond which was a device to give it an electric shock. If it turned to the right, all of these hazards were avoided. At first, the worms turned left about as often as they turned right. But the oftener the worm was made to crawl through the tube, the less often it did turn to the left, until after twenty to a hundred experiments, the worm learned to turn to the right and lead a happier existence. Evidently, the worm had slowly learned by trial and error that things were easier in the right-hand passage. The worm had learned!

Later other investigators repeated the experiment and found that if the front end of the educated worm containing the brain was removed, the worm, after growing a new head and brain, reacted as intelligently as before. Removal of the brain did not destroy the habit. The earthworm's ability to learn is apparently not centered in the brain as it is with the higher animals.

Why do sweets sometimes cause a toothache?

Sweets sometimes cause a toothache because of the ability of sugar solutions to extract water from the nerve of the tooth. This ability of certain solutions to draw water through a membrane is an example of *osmosis,* one of the most interesting yet mysterious effects found in nature. It is not only responsible for many toothaches, but it also plays a significant role in the process of life itself. Oxygen passes from the lungs to the blood, and carbon dioxide passes from the blood to the lungs by osmosis. Plants receive their nutrients from the soil by osmosis and the same process replaces the moisture in dried foods when they are soaked in water.

An interesting experiment is often used to demonstrate this phenomenon. A carrot is hollowed out and the cavity partly filled with a thick solution of sugar and water. The top of the cavity is then plugged with a cork through which passes a hollow glass tube. The carrot is then placed in a jar of water. It is soon noticed that water passes through the tissue of the carrot into the cavity and then up the glass tube. The height of the liquid in the tube is a measure of the osmotic pressure that has been developed. It is this pressure which is often responsible for toothaches. If the enamel is decayed around the tooth, the relatively porous dentine becomes the only barrier between the nerve and the sugar solution in the mouth. Osmotic pressure then comes into play producing an effect on the nerve which we interpret as pain.

The most satisfactory theory concerning the nature of osmosis is based on reduced mobility of molecules in a solution. When a substance is dissolved in water, there appears to be a union or binding between the water molecules and the molecules of the dissolved substance. These clusters of molecules have less freedom of move-

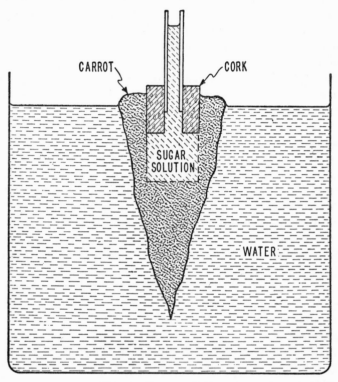

FIG. 4. OSMOTIC PRESSURE

Osmosis causes water to pass through the carrot into the sugar solution. This forces the solution to rise in the glass tube.

ment than the molecules of pure water and find it much more difficult to pass through the membrane or tissue. Water molecules, on the other hand, are smaller and more mobile and pass through the membrane with greater ease. Consequently, water will continue to pass through until the pressure on the other side is sufficient to equalize the rate of water diffusion in the two directions.

Why are modern airplanes pressurized?
At higher and higher altitudes, the pressure of the atmosphere gets progressively lower. This means that there are fewer molecules of oxygen to breathe. At 18,000 feet, for example, there are only half as many oxygen molecules in each cubic inch of air as there are at sea level. Naturally, less and less oxygen will enter the blood stream as one ascends to higher and higher altitudes. At 25,000 feet, the hemoglobin in the blood picks up only about half of its capacity of oxygen. At this altitude, unconsciousness may occur and, in some cases, even death.

The effect of a lack of oxygen is felt long before this altitude is reached, however. At 5,000 feet, breathing becomes more rapid. At about 12,000 feet, breathing is about twice as deep as at sea level. At 18,000 feet, vision, hearing and the ability to think clearly are seriously impaired. All of these symptoms develop without causing any alarm to the person affected. For this reason, all modern airliners are pressurized in order to provide transportation at elevated altitudes without the discomforts associated with these altitudes. A pressurized plane is designed and manufactured so that it is practically airtight. Rarefied air from the atmosphere is then pumped into the almost airtight interior. Enough air pressure is maintained in the cabin to avoid the low-pressure hazards mentioned above.

Why do roller coasters give us a thrill?
Energy can be made available to us in many forms. Chemical energy from a storage battery starts our car, solar or light energy heats the earth and atomic energy can destroy a city. Kinetic and potential energy are two forms which combine in the roller coaster to give us a thrill. Kinetic energy is energy of motion. An automobile, by virtue of its motion, is said to possess kinetic energy. The amount of this energy depends upon how fast an automobile moves and how heavy it is. The same automobile has no kinetic energy when it has come to a stop. If it has stopped on a hill, however, it may have a great deal of potential energy, the kind of energy contained in a wound clock spring. If we release the hand brake, the automobile will coast down the hill, gradually picking up speed as it goes. At the bottom of the hill all of the potential energy will have been converted to kinetic energy and the automobile will be moving at its maximum speed. This interchange of energy from one

37

form to another follows the law of conservation of energy. It is exactly this principle which enables a roller coaster to move up and down hills once it has been pulled to the top of the first long rise. Since there is always a certain amount of friction, the hills become progressively lower until the ride is terminated. The hills are designed in such a way that the train is barely moving at the top of each rise. This provides maximum change in speed between the top and bottom of each hill. The rapid change in speed toward the bottom of the downgrade causes that sickening feeling in the pit of your stomach. A pilot feels the same sensation as he pulls out of a steep dive. It is interesting to observe that speed alone cannot cause this sensation. It comes about, not because you are traveling so fast, but because your speed is *changing* so fast.

Why don't we ever see the back half of the moon?
The moon moves like a wrestler circling his opponent. Both the moon and the wrestler keep their faces toward the center as they rotate. For this reason, we *never see the back of the moon*. In scientific terms, the moon's period of rotation (about its axis) is precisely the same as its period of revolution (around the earth). It travels around the earth once in about $27\frac{1}{3}$ days and rotates about its axis in exactly the same length of time. This means that we always see the same half of the moon's surface. Actually this is only approximately true since the moon wobbles back and forth somewhat as it spins. This enables us to see 18 per cent more than half of its surface. The other portion of the moon, though nearer to us than any of the other celestial bodies, is completely unknown and inaccessible to our telescopes. We know more about Mars than we do about the other side of the moon.

How does sap rise from the roots to the leaves at the top of a tree?
The movement of water and dissolved nutrients often exceeds three hundred feet in such trees as the Douglas fir and redwood. The speed at which water rises varies from a negligible rate during humid weather to as much as three or four feet a minute on a warm summer day. How does an inanimate tree accomplish such a feat? In the human body, all parts are serviced by the blood stream, but there is no mystery about how it is done: each of us has a pump. Plants and trees do the same thing effortlessly, quietly and without

the use of mechanical gadgetry. How is it done? Scientists do not really know.

It seems highly probable that no one theory can fully account for the rise of sap under all conditions and at all times. Most scientists believe that several forces co-operate to produce this effect.

At one time it was thought that atmospheric pressure might be responsible but we now know that this can only account for a rise of about thirty feet. The theory that capillary action might accomplish this rise is also discarded today. Another theory states that root pressure, a force resulting from the absorptive activities of the root tissues, can account for the ascent of sap. This force can be measured by cutting a plant off at the base and attaching pressure measuring devices to the remaining portion. Pressures as high as eighty or ninety pounds per square inch have been measured in this way. This would be enough to account for the growth of trees, but for one objection: high root pressures do not occur when they should. Actively growing plants, which need large amounts of water display the lowest root pressures, while dormant, inactive plants have the highest root pressures. Root pressure is probably only part of the answer.

The most widely accepted theory states that transpiration and cohesion of water account for most of the rise of water in plants. Transpiration is the evaporation of water from leaves. Cohesion is the attraction of one water particle for another—the force which causes water to form droplets. This force has been estimated by physicists to be as great as fifteen hundred to three thousand pounds per square inch, much more than is necessary to account for our highest trees. Transpiration supplies the upward pull on many minute columns of water extending from the leaves down to the root hairs. As water evaporates from the cells of the leaf, it leaves a water deficit, or vacuum, in the cells directly beneath the surface. These draw upon the cells next below for a new supply and the process continues down to the root system of the plant.

Plants and trees are truly amazing. They live a complete life without a nervous system; they breathe without lungs; they distribute sap without a heart; and they are the only living things that can convert lifeless inorganic materials into the substances which all animals require for life. When we learn more about the secrets that plants have used for ages, we will know more about life itself.

Why is "bluing" added to detergents to whiten laundry?
Science tells us that ordinary light which we call "white" really contains a combination of all colors from violet to red. The sensation we call "color" is determined by the color of the light that happens to reach our eye. In reality, no object can have a color. The property which it does have is the ability to absorb light of a certain color and reflect others. If we somehow subtract red light from a beam of sunlight we find that the color of the resulting light is green. An object that looks green to our eye, therefore, must be able to absorb red light. Similarly, an object which looks yellow must have the property of absorbing blue-violet light. These pairs of colors, red and green, and yellow and blue-violet, are called complementary colors of light. There are many such pairs of complementary colors of light and all have the property of producing white light when added together in the correct proportions. When one color of the pair is absent for any reason, the resulting beam assumes the color of the complement.

The yellow which is sometimes present in laundered clothing is due to the absorption of blue-violet light by the material. The light reaching our eye has the normal amount of yellow but is deficient in blue-violet. The weak bluish dye contained in commercial bluing has the ability to absorb some of this yellow light to bring the amount of reflected yellow into balance with the blue. When this balance has been effected, the article again looks white.

How does an atomic battery operate?
An atomic battery converts nuclear energy directly into electrical current. The radioactive material used in its construction is strontium-90, which is produced as a by-product in nuclear power plants. Strontium-90 continually emits a few high-speed electrons in its process of disintegration into a more stable element. A plate of a second material, the metal silicon, is placed in such a position that some of these electrons strike its surface. For every high-speed electron that strikes a silicon atom, 200,000 low-speed electrons are released. It is the sum of all these low-speed electrons which produce the usable electric current. The power available from such batteries is low at the present time, but continued research will undoubtedly produce higher-powered units in the future. Since strontium-90 has such a long life, a battery using it would continue to deliver its

current for about twenty years. The first application of such atomic batteries may be in such products as hearing aids and small-sized radio receivers.

What does the octane number of a gasoline mean?

In normal operation, the internal combustion engine compresses a mixture of air and vaporized gasoline which is ignited at the correct instant by a spark plug. Some gasolines, when compressed too much, explode prematurely causing knocking and loss of power. Instead of a smooth, steady push, the top of the piston is given a sudden, hammer-like blow at the instant the fuel ignites. To get maximum power from the fuel, the mixture of gasoline vapor and air must be highly compressed. This causes the engine to knock badly unless a gasoline with good antiknock properties is used.

There are two ways by which the tendency of a gasoline to knock may be minimized: (1) by using a large percentage of slow-burning gasolines in the fuel; and (2) by using a catalyst to slow up the rate of combustion of the fuel. The best known of the catalysts is tetra-ethyl lead.

The term *octane number* is applied to gasolines to measure their tendency to knock in engines. A high octane number designates a gasoline with less tendency to knock than a gasoline with a low octane number. Iso-octane is the standard used for arriving at octane numbers. This gasoline is excellent in its antiknock properties. Another gasoline, n-heptane, has very poor antiknock characteristics. Pure iso-octane has arbitrarily been given an octane number of 100, while n-heptane has been given an octane number of 0. The octane number of any mixture of these two is equal to the percentage of iso-octane that it contains. Thus, a mixture containing 85 per cent iso-octane and 15 per cent n-heptane has an octane umber of 85. In order to determine the octane number of a given type of gasoline, it is burned in a standard engine and compared with various mixtures of iso-octane and n-heptane. The octane number of the gasoline under test is then taken to be the value of the iso-octane–n-heptane mixture which has the same knocking characteristics.

Good gasolines have a rating of 80 or higher. A 100-octane gasoline, such as is used in aviation, has as good antiknock characteristics as pure iso-octane. Recently, chemists have produced gasolines with even better anti-knock properties than pure iso-octane. It is possible, therefore, to produce a gasoline which surpasses 100-octane in performance.

Can insects think?
This subject has been clouded considerably as a result of the errors made by many early biologists. They wrongly ascribed to intelligence the many reflexive or instinctive actions performed by animals and insects. As a result, it has become difficult to look at the subject objectively and admit the existence of the thought process even where every indication proves it does exist. You would, for example, consider it an intelligent act to repair a broken beam before your porch fell down. The act of recognizing a structural weakness and effecting the necessary repairs surely requires the use of thought on your part. Let's see how the wasp reacts to a similar situation. Wasps normally attach their nest firmly to a suitable support. If the wind should loosen the nest and put it in danger of falling, the wasps repair the damage as soon as possible before returning to their normal activities. Also, consider the distribution of the work load within the wasp colony. Ordinarily, certain workers are devoted to gathering food, others to nest maintenance, others to brood tending, etc. Each worker seems to know what is expected of him and the work of the colony gets done smoothly. But if the nest should become damaged or get dried out due to the weather, the nest maintenance force will be aided by other workers that drop their normal activities to help in the repair work or carry water as may be required. If instinct were responsible for this sudden change in activity, we'd expect that all of the wasps would react in the same way. Since only a few change their job classifications we must assume that some sort of primitive thought process is involved.

The bee gives us an example of thoughtful communication which is even more striking. Bees normally concentrate on collecting the nectar of only one kind of flower at a time. It has been shown that a bee, upon finding a new type of flower in bloom, will make known to the others the fact that it has been discovered, that the nectar is plentiful, and that its location can be found by following the directions he is about to give. Amazing? It is all done through the so-called nectar dance. Upon returning to the hive with a good load of honey, a bee will dance up and down on the honeycomb to attract the attention of the other bees. The others immediately recognize the type of flower by the perfume emanating from the dancer. This stimulates them to seek nectar from the same type of flower. It has also been shown conclusively that the nectar dance indicates to the other bees the direction to take to get to the flowers from which the dancer has just returned. This is accomplished by orienting the angle of the dance with respect to the sun. The bees can then approximate this angle to the sun when flying from the hive and find the field of flowers described by the dancer. In the course of their everyday activities, therefore, bees communicate with each other concerning the quantity of food involved, its odor and its approximate location. Many entomologists find it difficult to ascribe this behavior to anything other than the use of the thought process on the part of the bee.

How does a tape recorder reproduce sound?

Sound is recorded on tape by means of magnetism. In magnetic recording, a narrow tape is coated with very finely divided particles of powdered iron. Although these particles are small, they exhibit all of the properties of magnetism that we find in larger pieces of iron. A constant-speed motor causes the tape to be moved past an electromagnet which in turn is connected electrically to a microphone. When sound strikes the diaphragm of the microphone, an electric current is produced which varies in intensity in the same manner as the sound waves. This electric current produces corresponding variations in the current flowing through the electromagnet. This, in turn, causes the magnetic strength of the electromagnet to vary in the same manner as the sound into the microphone. As the tape moves past the electromagnet, the variations in strength of the magnetism on the tape follow exactly the sound variations which produced them. When the tape is played back, its magnetic variations cause similar variations in the current produced in the electromagnet, which is then amplified and sent through a loudspeaker.

How do we distinguish between various musical instruments playing the same note?

A given note will have the same fundamental frequency whether it is played on a piano or on a saxophone, yet we have no difficulty in distinguishing between the two. A good musician can easily tell whether a note is played on a Stradivarius or an ordinary violin.

43

The difference between these sounds is a property called "quality." It results from the fact that musical instruments produce not only the fundamental frequency, but also a number of overtones. The so-called fundamental frequency determines the pitch of the note and is the same for all instruments. Overtones, as their name implies, are higher in frequency (or pitch) than the fundamental and are always exact multiple of the fundamental frequency. The first overtone is twice the fundamental frequency, the next is three times the fundamental, etc. The difference in quality between musical instruments is due completely to the difference in quantity of each of these overtones. The French horn, for example, consists almost entirely of the fundamental and the first overtone while the trumpet gets its tonal quality from a high percentage of the higher-order overtones. The human ear is capable of "measuring" the content of the various overtones in the sound and our brain uses past association to advise us of the sound's origin.

What causes earthquakes?

Strangely enough, earthquakes seem to begin with wind and rain. It is obvious that the eroding effect of wind and water would eventually wear down the mountain ranges almost to the level of the sea were it not for an opposing force which appears to push the mountains back up again. To illustrate, if we were to partly fill a hot-water bottle and press in at one point, it would merely bulge out at another. As rivers carry sediment down from the mountains and deposit it on the coastal regions, the continent becomes unbalanced. The pressure on the sea coast increases due to the increased weight of the sediment, while the pressure in the mountain regions is decreased. The result of this unbalance would be negligible if the rocks several miles deep in the earth were rigid and unyielding. It appears, however, that these rocks are subject to plastic flow much like the water in the hot-water bottle. As the coastal regions sink gradually under the ever increasing weight, the mountain ranges are slowly pushed up as erosion lightens them. This tends to keep the downward pressure about the same at both places. This slow bending of the surface crust of the earth results in the generation of tremendous forces until the strain is so great that the rocks break. The line of this break is called a fault and its formation is usually ac-

44

companied by shock waves called earthquakes. On the basis of this theory we would expect to find the worst earthquakes in those regions where mountain ranges are located near the sea, and such is usually the case.

Is the universe expanding?

Science tells us that distant stars are receding from the earth at extremely high rates of speed. The more distant the star from the earth, the faster it recedes from us. This startling bit of knowledge was deduced from measurements made on the wave lengths of light emitted by certain elements on distant nebulae.

Everyone is familiar with the reduction in pitch of a train whistle as the train passes by. As the train speeds away, the sound's wave length becomes longer, which results in the lower-pitched sound. The situation is analogous in the case of starlight. When scientists examine the light of these distant stars, they find that the wave lengths are always much longer than they would be if the stars were stationary. The more distant the star, the longer the wave lengths are found to be. Astronomers warn us, however, not to infer from these findings that the earth is the center of this expansion. They point out that our observations would be the same no matter where we were located in space. To illustrate, imagine a balloon, the surface of which has been covered with many small spots of paint. As the balloon is blown up, each spot recedes from every other spot at varying rates of speed. Any two spots which are close together will move apart slowly while distant spots will recede from each other at a higher speed. It is important to observe that any spot can be taken to represent the earth, with the same results.

The fastest-moving nebula measured to date is receding from the earth at the speed of 38,000 miles per second which is about one-fourth the speed of light. Astronomers estimate that this nebula is about 700,000,000 light years away from us. Using these figures and the fact that its speed will continue to increase as time goes on, we discover that it will double its distance from us in somewhat under 2,000,000,000 years. If we take into account the probable errors involved in the measurements, we may conclude that the universe expands at a rate such that any two nebulae double their separation every few billion years.

Can friction be eliminated by polishing surfaces?

When we attempt to slide one object over another we notice a force which tends to oppose the motion. This force is called friction. It is due to the surface irregularities of the objects being rubbed together. Among the many methods which have been devised to reduce friction are the use of: (1) polished bearings, (2) antifriction metals, (3) ball or roller bearings, and (4) lubricants. One would expect that any bearing surface could be improved by extremely fine polishing in order to prepare perfectly flat surfaces. Efforts in this direction have shown, however, that there is a limit to which friction can be reduced by polishing. If the surfaces are "too" smooth, the friction between them actually increases. Physicists believe that making surfaces too smooth enables the molecules of the surfaces to get too close together, thereby producing attractive forces similar to those which hold the molecules of any solid together. This so-called adhesion of one material for another results in the transfer of minute quantities of material from one surface to the other when they are rubbed together. In order to overcome this condition, automotive engineers reduce the friction between cylinder walls and pistons by making one surface *less* smooth than the other. Paradoxical as it may seem, this results in the most efficient design possible by reducing friction to a practical minimum.

Why is it easier to balance on a bicycle when it is moving than when it is stationary?

There are at least two factors involved in the successful operation of a bicycle. First of all, gyroscopic forces must be considered because of the rotating wheels. A gyroscope consists of a wheel free to spin on its axis within a light framework. When such a wheel is spinning, the frame will stay in the same plane unless considerable force is used to change its direction. Gyroscopic forces, therefore, tend to oppose any change in orientation of the bicycle. The second factor to be considered is centrifugal force. This is the force which pushes you against the side of the car when negotiating a sharp curve. As the bicyclist starts to fall ever so slightly, he turns the front wheel in the direction of the fall and centrifugal force pushes him upright again. The path of the bicycle curves first to the right and then to the left as the bicyclist compensates for each tendency to fall. It should be observed that turning the front wheel of a stationary bicycle does no good since centrifugal force can be produced only by motion in a circular path. It is a combination of these two forces which make it easy to balance a moving bicycle.

Why does it usually get warmer during a snowfall?

If we heat a pan of crushed ice while measuring the temperature of the ice and water mixture, we find that the temperature *remains constant* at o° C. until all the ice is melted. Although we are continually adding heat to the mixture, its temperature does not change. This heat is being used to change the solid ice to water. Since this heat results in no temperature change it is sometimes called latent (hidden) heat. Conversely, water at o° C. can be changed to ice at the same temperature by removing an amount of heat from it. The quantity of heat involved is much larger than one would expect. In changing from water to ice at the same temperature, a gram of water must give up eighty calories of heat. This is more than enough to raise that same gram of water from room temperature to the boiling point!

When snow is formed in the atmosphere, great quantities of heat are given up by the freezing water. It is this heat which causes the air temperature to rise slightly during a prolonged snowfall.

Why does the atomic bomb explode?

In 1939 two German scientists, Meitner and Hahn, performed an experiment which demonstrated a new and fantastic type of nuclear disintegration. All previously studied radioactive disintegrations involved the loss of a small particle (such as an electron) accompanied by the radiation of a relatively small amount of energy. Meitner and Hahn extended this study by bombarding an isotope of uranium (U-235) with neutrons, and found that the disintegrations resulted in elements with an atomic number around 45, several neutrons, and an unusually large amount of energy. The atom, or more properly the nucleus of the atom, had at last been smashed. A large amount of energy was liberated and several particles (neutrons) of the type that started the reaction were produced by it. This new type of nuclear reaction was called a fission reaction.

It was then clear to scientists of all nations that an atomic bomb could be made if sufficient uranium-235 could be obtained. To illustrate this requirement, let us imagine that we are preparing an atomic bomb. How would we go about it? First of all, we will need some uranium-235. Let's put a small amount, perhaps an ounce, in a container. If we investigate the disintegration of this uranium-235 we find that every once in a while one of the atoms explodes spon-

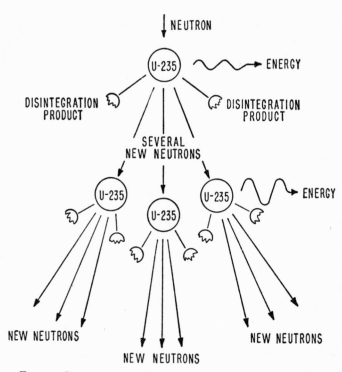

Fig. 5. Principle of the Uranium-235 Chain Reaction

A stray neutron strikes a nucleus, causing it to explode. This explosion, or atomic disintegration, produces nuclear fragments, energy and several new neutrons. Each of these may strike another nucleus and so on.

taneously into several smaller pieces. When this happens, the nucleus breaks up and in the process a few neutrons excape. They travel at low speed until they either escape entirely from the uranium-235 or strike another atom of the material and cause it to explode. The chances of striking another atom of uranium-235 are not as good as you would suspect at first glance. This is because most of the atom is really empty space. There are extremely great distances, relatively speaking, between the small particles that make up an atom. As a result, we would find that most of the neutrons developed by our small atomic explosions fly through the uranium-235 without hitting another nucleus, and are lost. As we add additional uranium-235 to our bomb, however, we increase the chances of secondary explosions because the neutrons must travel through a greater distance of uranium-235 before they can leave the material. Scientists have shown that there is a critical mass of uranium-235 required for a self-sustained chain reaction. Above this mass, each atomic explosion will produce at least one more and so on until most or all of the atoms have exploded. Below this mass, the reaction will die out. The critical amount of material for such a chain reaction is about twenty pounds since this amount provides a sufficient number of targets for the first few secondary neutrons.

Naturally, we would not want to put twenty pounds of uranium-235 in a pile until we wanted the bomb to explode. The firing mechanism must do this for us when the bomb is "on target" and we are elsewhere. We would probably divide the material into several equal parts and separate these parts to prevent the neutrons of one part from acting on the atoms in another. When we want to explode the bomb, all that is necessary is to cause the several parts to be pushed together rapidly by a suitable mechanical device. It is interesting to observe that only about one-thousandth of the mass of U-235 is converted into energy in the atom bomb. This amount is sufficient, however, to provide the bomb with the destructive power of twenty thousand tons of TNT.

What is thunder?
Thunder is merely a secondary effect caused by lightning. It results from air rushing into the vacuum caused by the bolt of lightning. It usually arrives at the observer some time after the flash is seen because sound travels much more slowly than light. It is useless to fear thunder, because by the time its sound reaches an observer, the bolt of electricity has already done its work. As a matter of fact, there is little reason for people to be afraid of lightning itself. The deaths caused by lightning in the United States amount to only

one per cent of the deaths caused by automobiles. As science learns more about lightning, this figure will undoubtedly be reduced. Perhaps we will someday learn to do the same for the automobile.

Does all wood float?

There is great variety in the density of the various kinds of wood. The logs used to construct the raft used in the *Kon-Tiki* adventure were made of balsa, which weighs only 6 pounds per cubic foot. Among the seasoned woods used for building and furniture in this country, the density seems to vary from about 22 to about 51 pounds per cubic foot. White pine weighs 25, maple weighs 44, and hickory weighs 51 pounds per cubic foot. Since fresh water weighs 62.5 pounds per cubic foot, most seasoned wood used in this country will float. The heaviest wood found in the United States is the black ironwood of Florida, which weighs 65 pounds per cubic foot. This wood must sink since it cannot displace its own weight of water. But the heaviest woods are to be found in other parts of the world. One of these is the poison ash of Mexico, weighing 69 pounds per cubic foot. The acapú, the ipé and the arapoca, which grow in the Amazon jungles, weigh respectively 70, 73 and 75 pounds. Various species of ironwood and lignum vitae grown in Puerto Rico reach a density of 85 pounds per cubic foot. The quebracho, or "Ax Breaker," of Argentina reaches an amazing 87 pounds, as does the West Indian ebony. These are the heaviest of the known trees. Since wood weighing more than 62.5 pounds per cubic foot will sink, it can be seen that many kinds of wood cannot possibly float.

Do wrecked ships stop sinking when a certain depth is reached?

While many find it hard to believe, wrecked ships will sink to the bottom of the ocean, even at its deepest point. We owe this bit of knowledge to Archimedes, the famous Greek physicist and mathematician who lived in the third century B.C. As a result of his experiments he discovered that a submerged body is buoyed up with a force equal to the weight of the water displaced. An object having a volume of one cubic foot will displace exactly one cubic foot of water. Since the weight of this amount of water is 62½ pounds, the object is buoyed up by 62½ pounds. Stated another way, a submerged object experiences a weight reduction which is equal to the weight of the water which it displaces. So long as a submerged object has any weight left at all, it will sink. For a wrecked ship to stop sinking, it much reach a depth at which it can displace its own weight of water. This might happen if water were compressible and became more dense and therefore, heavier, at the lower depths. Such is not the case, however, as water is very nearly incompressible. It must be concluded, therefore, that ships must continue to sink until they reach the bottom of the sea.

How does the Rh factor affect childbirth?

To discuss the Rh factor, we must first understand the nature of antibodies and antigens. An *antibody* is a chemical substance produced by the body to help ward off an invasion by a foreign substance called an *antigen*. Antibodies are never present in the body of an individual until required by the unwelcome presence of an antigen. An antigen may be one of a multitude of different kinds of substances present in an invading microbe, but the resulting antibody affects only that specific antigen. In this way, the body seems able to ward off the attacks of foreign antigens by the production of specific antibodies.

There is a great deal of discussion today of the Rh blood factor. This factor is, in reality, an antigen which is present in the red blood cells of about 85 per cent of the caucasoid stock. The name "Rh" was given to this antigen because its presence was first detected in the red blood cells of the rhesus monkey. Blood containing the Rh antigen is called Rh-positive, while blood free of this antigen is called Rh-negative. If Rh-positive blood, containing the antigen, is transferred into the blood stream of an Rh-negative person, that person's body is stimulated to produce opposing antibodies. These antibodies attack the Rh antigen in the red blood cells so introduced and in the process, the red blood cells themselves are destroyed. These antibodies persist, as if by habit, even after all the Rh-bearing red cells are destroyed. We then say that the person has become sensitive to the Rh factor.

A person who is Rh-positive does not produce an antibody to oppose the Rh antigen in his blood since the Rh factor is normal in his blood. If the blood of a sensitized person were transfused into such an individual, the Rh antigen would be attacked by the antibodies in the donor's blood. This would result in the destruction of the recipient's red blood cells. To prevent this, hospitals not only determine the blood type, but they also test for the Rh factor before giving a transfusion.

If an Rh-negative woman (no antigens) is bearing an Rh-positive child, difficulties sometimes occur. If any of the child's red blood cells, which contain the antigen, get into the mother's blood system, the antigen will stimulate the production of the antibodies. These, in turn, can go through the placenta and get into the blood of the infant. When this happens, the antibodies destroy the child's red blood cells which contain the antigen. The damage may be so severe as to cause death and abortion. In less severe cases, the child is born with a serious case of jaundice, due to the breakdown products of the red blood cells. Such jaundiced children may be saved by prompt and frequent blood transfusions during the first few months after birth.

In many cases of Rh-negative mothers bearing Rh-positive children, no problems arise. If such a mother has never been sensitized to the Rh factor, the pregnancy may only result in making her sensitive and not in serious illness of the child. Therefore, the first such pregnancy usually goes through normally. In subsequent pregnancies, she may still be able to have normal children if their births are spaced widely enough apart to allow the concentration of antibodies in her system to fall to a low enough level before conception occurs.

If the mother is Rh-positive, no problem arises since she cannot produce antibodies for transmittal to the child. If both parents are Rh-negative, they cannot have an Rh-positive child, so the problem does not exist. Even if the father is Rh-positive, it is possible for the mother to bear Rh-negative children, which are, of course, not liable to infantile jaundice.

The importance of this factor to a mother and her child has served to emphasize the importance of the early typing of the blood of female children. If an Rh-negative girl is given a blood transfusion from an Rh-positive person, she may become sensitive to the factor as discussed above. This might make it almost impossible for her ever to bear an Rh-positive child. In addition, if she should ever receive a second transfusion of Rh-positive blood, the antibodies present in her system would attack the new red blood cells resulting in a dangerous situation. For this reason, every woman should be typed for the Rh factor at an early age, and definitely before marriage.

Does a baseball really curve?

Yes. The fact that a baseball curves is based on an effect discovered over two hundred years ago by Daniel Bernoulli, the famous Swiss mathematician. The Bernoulli effect, as it is known, states briefly that the static pressure of a fluid, such as air or water, goes down as its velocity goes up. On the surface, it doesn't appear to make sense. Bernoulli tells us, in effect, that the pressure within a hurricane

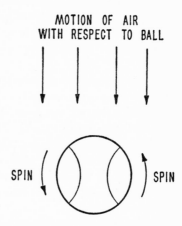

MOTION OF AIR
WITH RESPECT TO BALL

SPIN   SPIN

FIG. 6. THE CURVE OF A BASEBALL
When a thrown baseball spins, air is carried around with it by friction. This moving air adds to the air speed on one side and subtracts from it on the other. The ball will move in the direction of the higher-speed air—to the left in this example.

is less than the pressure in our living room. He resolved this apparent paradox by pointing out that there are two kinds of pressure, dynamic and static. Dynamic pressure results when a moving fluid (a liquid or gas) comes into contact with an object and it is this kind of pressure which is responsible for the damage done by high winds. Static pressure, on the other hand, is the pressure existing within the fluid itself, the pressure of one atom against another. While dynamic pressure goes up with velocity, static pressure goes down in the same proportion. But what does all this have to do with a 3-2 pitch in the last half of the ninth inning? Everything, if the catcher signals for a curve ball. As the pitcher releases the baseball, he imparts a spinning motion to it which causes it to rotate about a vertical axis on its path plateward. As the ball whirls, air is carried around with it by friction. On one side of the ball this air moves

53

with the current of air caused by the forward motion of the ball, while on the other side it moves in *opposition* to it. This causes the air speed on one side of the ball to be greater than on the other. The ball must curve, therefore, toward the side having the greater air speed and, consequently, the lower static pressure.

It has been proven experimentally that a baseball really does curve in accordance with Bernoulli's theorem by throwing a ball through a series of parallel screens made of fine mesh. By measuring the path of the ball in this manner, it has been demonstrated that a baseball can be made to curve as much as 6½ inches from its normal path.

Attempts have been made photographically to prove that this whole affair is in reality an optical illusion. In evaluating such evidence, it should be remembered that a camera is also subject to optical illusions unless handled in a scientific manner. In any event, the subject has long been closed to most baseball players.

If you add one quart of water to one quart of alcohol, do you have two quarts of liquid?
Most automobile drivers have, at one time or another, been led to the conclusion that two objects cannot occupy the same space at the same time—especially if they are automobiles. Along similar lines, scotch cannot be poured into a decanter unless we allow air inside the decanter to escape. A good carpenter will drill a hole before putting a screw into a piece of wood to prevent splitting. All of these examples illustrate the property of matter which is called impenetrability. Actually, two objects can sometimes occupy the same space at the same time if one of them happens to be porous. The skeleton of the sponge is a typical example of a porous solid. Somewhat less obvious is the cement block used in home construction which must be waterproofed on the outside to keep the basement dry. Metals such as silver and iron seem reasonably solid but water under high pressure can be forced even through them. When we consider liquids, there would seem to be little possibility for porosity, but water, nevertheless, does seem to be somewhat porous. If we mix one quart of water with one quart of alcohol, we find the resulting mixture adds up to slightly less than the two quarts we would expect. This is due to a slight porosity on the part of both liquids. Even though they are almost incompressible, both liquids have

spaces between their molecules. Some of the water molecules slip in among the alcohol molecules, and some of the alcohol molecules find room between the water molecules. This action results in a slight but measurable reduction in the expected volume of the mixture.

How much work is performed in pulling on a stuck drawer?
Although pulling on a stuck drawer may be an exasperating and tiring experience, it does not involve the accomplishment of work, at least in a scientific sense. Most of us think of studying, holding a bag of groceries, or even writing a letter as forms of work. To the scientist, work is a measurable physical quantity in the same sense as length and weight. To qualify as a form of work, a force must push an object through a distance. Since the stuck drawer is presumably motionless, no work can be done upon it no matter how much muscular effort is involved. In the same sense, no work is expended in attempting to push an automobile out of the mud unless it moves. Work can be accomplished, however, in pushing an automobile up a hill, throwing a baseball, or carrying a bag of groceries up the porch steps. Note, however, that no work is done in carrying a sack of flour across a level room. If we drag it along the floor, on the other hand, we do work in overcoming the force of friction between the sack of flour and the floor. The test for work, then, is the existence of a force which operates on a body causing that body to move through a distance.

Why do matches light?
Two kinds of matches are in common use today; (1) the friction match, which can be ignited on any rough surface, and (2) the safety match, which can be ignited by rubbing it on a prepared surface. The friction match is made by first dipping the match stick into an ammonium phosphate solution. This prevents afterglow, and lessens the danger of producing fire as a result of careless handling of supposedly extinguished matches. Next, the head of the match is dipped into melted paraffin, after which it is dipped into a paste containing glue, lead oxide and a compound of phosphorus. Friction causes the phosphorus and lead compounds to explode and set fire to the paraffin, and this in turn sets fire to the wood.

In the safety match, the tip is composed of easily combustible antimony sulfide, and potassium chlorate, which provides additional oxygen to aid in producing combustion. The side of the box contains red phosphorus, a nonpoisonous form of the element. The material on the tip of the match will not ignite readily unless it is rubbed on the treated side of the box. The friction produced by rubbing the match against this surface vaporizes a trace of the red phosphorus, which ignites and sets fire to the head of the match.

What would life be like on the moon?
If one of the fastest jet planes were at our disposal, a three- or four-month trip would find us exploring the surface of the moon. Although we would see many mountain ranges and valleys, we would find no lakes or seas. In fact, there is no water whatever on the moon, nor is there an atmosphere surrounding it. This is because of the low gravitational pull of the moon (about 17 per cent of that of the earth), which is not sufficient to prevent the swiftly moving molecules of air from escaping into space. Its surface is an arid wasteland, lacking the protection of clouds or atmosphere which insulate us on earth from the burning rays of the sun during the day, and the rapid escape of heat at night. In addition to this lack of insulation, a day on the moon lasts almost thirty times as long as one on earth, permitting the noonday temperature to reach a sizzling 230° F. while the evening temperature falls below that of dry ice.

We would require the use of some sort of space-suit to protect us from the elements and provide the oxygen we need for life. The average man would weigh only about twenty-five or thirty pounds and would be able to jump about twenty-five or thirty feet straight up! In our wanderings about the satellite, we would note that the area of the moon is about ten times that of the United States or about thirty million square miles, consisting mostly of mountains and plains of volcanic origin. Nowhere on this vast desert would we find a single piece of vegetation or sign of life to remind us of our pleasant existence on earth. Indeed, astronomers have searched with powerful telescopes for a single town or large building, which ought to be visible if they exist. None have been found. We must conclude that life on the moon would be moderately interesting for a while, but extremely uncomfortable as a steady diet.

Why does a quarter fall faster than a feather?

You will recall that Galileo took time out from his many pursuits to perform his famous experiment of dropping stones of different size from the Leaning Tower of Pisa. Legend tells us that although the stones smashed to pieces at the same instant, the people of Pisa refused to believe their own eyes. Actually, Galileo must have cheated a little bit to make the stones hit at the same instant. Wind resistance should have caused the lighter stone to lag a little bit behind the heavier one. It is for this reason that a quarter falls much faster in air than a feather. Suppose you were to take two balls of the same size, one made of lead and the other of wood, and tried the same experiment over again. You would find that the lead ball falls a little bit faster than the wooden one. After falling a long distance, perhaps a mile or so, each would reach its own terminal velocity at which the resistance of air just balances the force of gravity. This velocity would be a little higher for lead than for wood. In the case of the feather, the terminal velocity is very low and is reached in short order after falling only a few inches. If we investigate falling objects in a vacuum, however, all of this changes. In a vacuum, all objects fall at exactly the same speed no matter what their shape or their weight.

Why does a gun kick back when it is fired?

Each time we pull the trigger of a gun we send a bullet hurtling through the air at tremendous speed. We might ask, "What pushed the bullet?" If we are told that the powder charge did it, then what did the powder charge brace itself against to effect this enormous push? The only answer is the gun. The gunpowder, in exploding, pushes just as hard against the gun, and consequently your hand or shoulder, as it does against the bullet. The bullet travels much faster than the gun because of the great difference in weight of the two items. If a gun weighs one thousand times as much as the bullet it fires, the bullet will travel one thousand times as fast as the gun after discharge. To put it scientifically, the weight of the gun times the velocity of the gun must equal the product of the weight and velocity of the bullet. There are some interesting applications of this rule in everyday life. If a billiard ball is struck a direct blow by another, the first assumes the velocity of the second. Take the case of driving a nail into a beam. If the beam is massive, the nail glides in smoothly while a nail driven into a flimsy beam will cause it to shake vigorously in response to the blows. Similarly, the strong man in the circus can withstand a blow from a sledgehammer by placing a heavy slab of concrete on his chest. The weight of the concrete is so much greater than the weight of the sledgehammer that its velocity of recoil is very low.

How many stars are there in the sky?

Ptolemy, one of our earliest astronomers, succeeded in counting about a thousand stars which he classified by brightness into six groups or, as he called them, magnitudes. Group 1 consisted of about twenty of the brightest stars, while group 6 consisted of those stars which were just visible to the unaided eye. Modern astronomers have retained this arbitrary scale but with the help of powerful telescopes have extended the range to the twentieth magnitude. Today we know that there are about five or six thousand stars visible to the unaided eye and about one billion or so if we add up all of the stars to the twentieth magnitude. This figure, by the way, must be very low because there are many clouds of dark matter in the sky which must hide many stars from our view.

Up to this point we are on firm ground in our star count since astronomers have scientific evidence that at least a billion stars do in fact exist. From here on, however, the problem gets somewhat "nebulous." An exact analysis must be left to scientists but considerable evidence exists to indicate that there are no stars beyond the twenty-sixth magnitude. If this is true, it is then possible to estimate the number of stars in each of the six unseen magnitudes by a method mathematicians call extrapolation. This is merely an educated guess made on the basis of the number of stars in each of the known magnitudes. In any event, calculations tell us that there must be at least thirty billion stars in the heavens. The actual count will probably be very much higher, however, when all the counting is completed.

How does a firefly light up?

While most of us take such ordinary phenomena as the incandescent lamp, the sun and the aurora borealis quite for granted, we seem to have a real curiosity about our little beetle friend, the firefly. Yet, if we study his light, we find that it is very much like other kinds of light. It can be measured, reflected, refracted, polarized, and yet it is produced without heat. Light such as this, which is produced without any heat, is called luminescence.

Luminescence, in the firefly, is produced by a substance called *luciferin* which combines with oxygen in the light-producing reaction. This reaction will take place, however, only if another substance, *luciferase,* is present. Luciferase acts as a catalyst which encourages but does not take part in the chemical reaction. This light-producing reaction can be made to take place outside of the body of the firefly by extracting these two substances and mixing them in a test tube. Scientists believe that the production of light in the firefly is not a vital process but merely a coincidental by-product of a chemical reaction taking place in his body. As a matter of fact, certain seagoing protozoa eject luciferin and luciferase into the surrounding water thus producing light outside the body.

The amount of light emitted by the firefly is not very great, equal to only about one-thousandth of the light of an ordinary candle. Although this light can be produced in the laboratory, the ingredients required must be obtained from the firefly. Chemists have not yet learned how to synthesize them and their secret still belongs to nature. But why nature should bother with animal luminescence in the first place is the real mystery. In some cases it seems to be associated with mating or the procurement of food, but for the most part its real purpose, if any, is unknown.

Why is ice slippery?

For thousands of years man has recognized that ice is slippery and has put this fact to use in simplifying his transportation problems. An automobile skids and a sled glides over ice because of the extremely low friction that ice affords to a moving body. But what is friction, and why does ice seem to have less than its fair share of it? If we examine any ordinary substance such as wood or steel under a microscope, we find tiny irregularities in the surface. When any two such substances are rubbed together these irregularities interlock and impede the rubbing motion. If a small amount of oil is placed between two plates of steel, we find that the friction is greatly reduced because oil prevents the plates from coming into direct contact with each other. Since the surface irregularities are separated to some extent, there is less tendency for them to interlock, and friction is reduced.

In the early part of the nineteenth century, Sir Humphry Davy found that he could melt ice by rubbing two cakes of it together. Aside from the astounding discovery that mechanical work and heat were interchangeable, this simple experiment explained to the world why it had been slipping on ice for countless ages. The simple

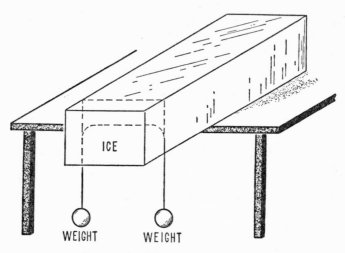

FIG. 7.   ICE MELTS UNDER PRESSURE

The wire will slowly pass through the slab of ice without cutting it in two. The weights exert pressure which melts the ice. As the wire falls, the pressure above it is relieved and the water turns back into ice.

truth is that ice melts under the pressure of an ice skate and the thin film of water between the runner and the surface of the ice acts as an excellent lubricant. As soon as the runner passes by, of course, the water changes back to ice. This is, no doubt, the reason that it took so long for man to gain an insight into the slippery nature of ice.

What is the theory of evolution?
The theory of evolution states that animals and plants have been undergoing continual and gradual change into new and different species since the infancy of our planet, perhaps four billion years

ago. In addition, such changes are held to be responsible for the various species in existence today and these species, through evolution, will provide the new forms of the future. Evolution is continually going on about us today, although its rate is much too slow to be observed in a generation or two. The important concept to gain from a study of evolution is that all living things are continually changing. These changes, however, are not necessarily for the better. While some organisms have improved through evolution, others have sunk progressively to lower and lower forms. In some cases evolution has led species up blind alleys which led to supremacy for a while only to be followed by eventual extinction. A typical example is given by the dinosaurs which ruled the earth about a hundred million years ago. A number of factors contributed to the decline of the dinosaurs. Since they were reptiles and consequently cold-blooded (their body temperature changed with changing air temperature), the slowly cooling climate made them sluggish. This cooler and drier climate also reduced the luxuriant vegetation which they needed to nourish their gigantic bodies. On top of all this, they had such small brains that they must have been remarkably stupid. The dinosaur was very well suited to the environment of its time but was physically and mentally unable to adjust to any serious change in that environment. The dinosaur and other reptiles were followed by the warm-blooded animals which began to displace them in importance. Although these animals were much smaller than the dinosaurs, they had relatively large brains. This permitted them to win the battle of wits even to the point of dining on dinosaur eggs which were left in convenient and unguarded places.

Another aspect of evolution is the law of survival of the fittest. The conger eel, for example, lays about fifteen million eggs in one season. We should be thankful that only a very small percentage of these eggs ever mature. Otherwise, we would have nothing but conger eels in the world. Presumably, those which survive must differ in some favorable respects from their less fortunate relatives. If this process were to continue for millions of years, a species ought to evolve which is better suited to its environment than its predecessor. Experiments have produced DDT-resistant strains of insects by just this means.

It's beyond the scope of this book to prove or disprove the theory of evolution but the weight of scientific evidence leaves little doubt

61

of its validity. Evolution does not attempt to explain the origin of life. It is a scientific theory based upon careful observation and experimentation which attempts to show how living things have arrived at their present state of development.

How old is the earth?
Several radioactive elements, such as uranium, undergo continuous disintegration into lighter elements and these, in turn, decompose into still lighter elements until only nonradioactive lead is left. It has been determined that the half-life period of uranium is 4,670,-000,000 years which means that if one were to start today with one pound of uranium, he would have only one-half pound left at the end of that period of time. The missing uranium would have disintegrated successively into such elements as uranium-234, ionium and radium and most of it would have changed ultimately to lead. Since lead does not decompose, it must accumulate, and from a knowledge of the half-life periods of the elements in the uranium-to-lead series, we can calculate the length of time necessary to reach the ratio of uranium to lead existing at present in a given mineral. This tells us the age of any rock which contains these elements and many such deposits have been found. By interpreting lead-uranium ratios in this manner, the earth is estimated by various authorities to have a minimum age of from two to four billion years.

Why are the eyes of small animals proportionately larger than our own?
The back of the eye, upon which an image of the outside world is focused, is composed of a mosaic of rods and cones, the diameters of which are approximately equal to the wave length of light. The human eye contains about a half-million of these light-sensitive elements. In order for two objects to be distinguishable, their images must fall on separate elements. It is obvious, therefore, that with fewer rods and cones we would see less distinctly. If the size of the rods and cones were diminished, science tells us that they would not respond to visible light at all. The eyes of small animals must, therefore, have rods and cones which are not much smaller than our own. The quantity of these elements depends, therefore, upon the size of the eye. In order for these eyes to be of any use at all, the eyes of small animals must be much larger in proportion to their

62

bodies than our own. Even with his relatively large eyes, a mouse finds it difficult to distinguish between two baseballs at a distance of three feet. Larger animals, on the other hand, require only relatively small eyes and those of the elephant are only slightly larger than those of man.

Why does water flow?

According to present-day theory, gases are composed of small particles or molecules which are in rapid motion in what would otherwise be a large empty space. As a matter of fact, most of the volume occupied by a gas under normal conditions is *empty space* and the individual molecules of the gas are separated from each other by relatively large distances. By virtue of their temperature, these molecules possess a certain amount of kinetic energy, or energy of motion. This energy causes them to move in straight lines until they collide with other molecules, whereupon they rebound like rubber balls on their way toward new collisions. In addition to their motion, molecules have a force of mutual attraction similar to the force of gravitation between the earth and the moon. If the kinetic energy of a gas is reduced to a low enough value and if the molecules are pressed closely enough together, this attractive force becomes sufficient to overcome the kinetic energy and the molecules come into direct contact with one another. The gas has then become a liquid. The molecules of a liquid are in contact but are able to slide over one another like a number of marbles in a bag. It is this mobility which distinguishes liquids from solids. As a liquid is cooled further, the kinetic energy is reduced to the freezing point which further limits the mobility of the molecule. Molecules in the solid state are believed to be so close together that their movement is restricted to back-and-forth motions about fixed points of equilibrium.

How long will the sun continue to shine?

We now know that the sun will continue to shine for billions of years in the future. Until the middle of the ninteenth century, however, the sun was presumed to be a body which was once much hotter and was in the process of slowly cooling off. Until this time, scientists did not realize that energy, as well as mass, is a measurable quantity and must, therefore, have an explanation for its origin and existence. They then began to reason that the sun cannot be simply burning for then it could have lasted only several thousands of years, even if it were composed of pure carbon and oxygen. Recently, almost conclusive evidence has been found to indicate that the sun's radiant energy is the result of atomic transformations of its elements in accordance with Einstein's principle concerning the equivalence of mass and energy. The quantities of energy involved in these transformations are of a vastly higher order of magnitude than ordinary combustion. This fact has been adequately demonsti ted by the destructive power of the atom and hydrogen bombs.

If the sun were to radiate energy at its present rate for the next 150,000,000,000 years, its mass would not be reduced by as much as one per cent. Consequently, it is not surprising to find that geologists believe that the earth has received energy at substantially the present rate for several billions of years, and will continue to do so for many billions of years to come.

What are precious stones made of?
A precious stone is actually a mineral having a certain form and chemical composition. Since ancient times, men have distinguished between ordinary minerals and those whose color and brilliance set them apart from the rest. Color in precious stones may be of two kinds: inherent or the result of impurities. Turquoise derives its lovely blue color from the material of which it is composed, copper aluminum phosphate. On the other hand, some minerals like quartz are colorless in the pure form but may sometimes be colored by small amounts of impurities. Strange as it may seem, most precious stones derive their value from a small amount of impurity. It is often the accident of nature that seems to appeal most to man.

Among the very precious stones, the ruby and sapphire have achieved the height of success. They are variations of the same mineral, corundum, the chemical name of which is aluminum oxide. These are the same aluminum and oxygen, by the way, that make up 56.7 per cent of the earth's crust. Corundum makes a good basic material for gems because of its great brilliance and because of its hardness, which offers lasting protection against scratching. If a trace of titanium (one of the ten most plentiful elements) happens to have found its way into corundum, the stone is colored blue and its value increases. Iron oxide imparts a yellow color to corundum and raises its value almost to that of the diamond. If the impurity happens to be chromium, the value of the stone is raised to such a point that it is given a name all its own and we call it a ruby. The difference between the value of the abrasive in sandpaper, corundum, and a carat of the same material with 3 per cent of chromium thrown in is about one thousand dollars.

Emeralds, which are extremely rare and valuable, are made of beryllium aluminum silicate. This mineral is not particularly attractive in itself, because of its lack of brilliance and hardness, but if it is colored green, it becomes extremely rare and expensive. The material that gives the emerald its beautiful deep green color is once again chromium, this time in the form of an oxide.

The major exception to the rule seems to be the diamond, which is considered most beautiful in the pure form. This is probably due to the inherent brilliance and fire of the properly cut stone. The diamond, as you know, is a form of pure carbon which occurs in many colored forms in nature. The blue-white form is most popular while the yellow and brown stones are not valued highly as gems.

Because of its outstanding hardness, the less valuable forms of the diamond are used by industry in the manufacture of cutting tools. The United States imports more diamonds for this purpose than for use as precious stones. Of course, the value of these industrial stones is considerably lower, but it is interesting to find that the most precious stone of them all is no longer just an ornament. The king of gems has been put to work.

What is a virus?

Modern theory tells us that a single virus is a highly complex protein molecule. Although these molecules are among the heaviest yet found and are many millions of times heavier than a hydrogen atom, they defy investigation because of their small size. They cannot be filtered and they do not reproduce themselves in the laboratory as do bacteria. Their isolation has been accomplished, however, by fractional distillation, a process for separating constituents of a mixture by taking advantage of their different boiling points. Such experiments seem to indicate that a virus is not "alive" in the normal sense, since ordinary organisms cannot undergo such an ordeal and remain alive. And yet, when the tobacco mosaic virus, for example, comes in contact with the tobacco plant, the molecules begin to multiply producing an amount of virus many millions of times that of the original. This ability to reproduce itself at the expense of its environment is one accepted characteristic of living things. Consequently, it has been proposed by some scientists that the virus exhibits a dual type of existence, alive and yet not alive; animated by a specific environment but passive in all other environments.

What are the inside facts about the earth?

Although our actual experience in penetrating into the earth is limited to about two miles, much has been discovered concerning the inside of this world of ours. We know, for example, that its radius is about 4,000 miles and its average density is about 5½ times that of water. Seismologists tell us that there is an inner core with a radius of about 2,100 miles, which behaves like an alloy of iron and nickel, having an average density of about nine times that of water. Surrounding this metallic core is a shell about 1,100 miles thick probably consisting of metallic iron and silicates having a density of about 5.7. Next comes a shell about 700 miles thick having an average density of about 3.5 and probably consisting of silicates of iron and magnesium. Finally, there is a shell about 50 miles thick which forms the surface of the earth. The vast interior of the earth seems to be very hot and all deep mines show a steady increase in temperature of about 1° F. for each 100 feet of depth.

There is a striking inequality in the distribution of the various elements making up the earth's crust. Oxygen makes up about one-half of the weight of our land, about one-fifth of the weight of our atmosphere, and eight-tenths the weight of the sea. The metal silicon constitutes more than one-quarter of the weight of the earth's crust. About 99 per cent of the earth's crust is made up of only twelve elements. It is interesting to note that the relatively uncommon element titanium is included among these abundant elements, while the copper of our everyday penny is absent.

The weight of the earth is 7,000 million million million tons and its cubic content is sufficient to provide each inhabitant of the United States with 1,700 cubic miles. That, you will agree, is quite a parcel of land.

Why does water wet glass while mercury will not?

Anyone who has ever tried it knows that walking in deep mud can be a tiring occupation. Similarly, a finger stuck into thick molasses can be withdrawn only with effort. A fly finds it impossible to extricate itself from the sticky surface of flypaper. These and similar situations arise from two characteristics of liquids: (1) the tendency of liquid molecules to stick to themselves, called *cohesion,* and (2) the ten-

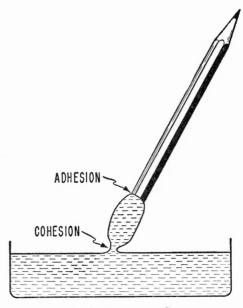

ADHESION

COHESION

Fig. 8. Adhesion and Cohesion
Adhesion and cohesion combine to prevent
the molasses from leaving the pencil.

dency of liquid particles to stick to other substances, called *adhesion.* When a finger is withdrawn from molasses, it is completely covered or wet with molasses. This is because of adhesion between molasses and finger. Pulling it out requires the use of a certain amount of force. This is because of cohesion of one molasses particle for another. It takes the expenditure of force to pull these particles apart. If we withdraw glass from water, we find that the glass is wet. This is due to the fact that the adhesion of water to glass is actually greater than the cohesive force tending to hold the water particles together. In the case of mercury, this is not true. The cohesive force of mercury is greater than its tendency to adhere to glass. Mercury, therefore, will not wet glass.

How does a radio tube work?
In 1883, Edison experimented with an electric light bulb to which he had added an additional metal plate. This plate was located within the bulb at a short distance from the heated filament and was connected to a wire which passed through the glass. The wire, in turn, was connected externally to the heated filament. Edison noticed that, under these conditions, a current of electricity circulated from plate through the external wire to the filament of his experimental tube. Since an electric current must flow in a closed path or loop, Edison reasoned that the electrons, which make up this current, must jump across the gap from filament to plate in the evacuated tube. It was later found that placing a positive voltage on the plate greatly increased this current flow by helping to attract the negatively charged electrons from the filament. In the light of present knowledge, this current is explained by the passage of electrons from the interior of the filament into the space surrounding it and the attraction of these emitted electrons by the positively charged plate. The emission of electrons from a heated filament is analogous to the evaporation of molecules from a liquid. In the case of electron emission, the filament must be heated to a temperature that provides the electrons with sufficient energy to overcome the force tending to keep them in the metal.

To this two-element vacuum tube, Lee De Forest, an American inventor, conceived the idea of adding a third electrode, or grid. The grid electrode consists of a fine wire mesh or screen which is located between the filament and plate but insulated from both. This electrode is connected to a wire passing through the glass in the same manner as the plate. The purpose of the grid is to act as a valve which, opening and closing, controls the number of electrons that are permitted to pass from filament to plate. As a matter of fact, British scientists use the more logical term "valve" to describe what we in America call a "vacuum tube." The way the grid performs this function is tied in with the fact that negative charges repel one another. If a strong negative charge is placed on

68

the grid, all of the electrons emitted by the heated filament will be repelled by the grid and will collect in a group around the filament. Eventually an equilibrium condition will exist wherein the number of electrons in this group will become constant. As new electrons are emitted from the filament, an equal number pop back into it again. Now if the negative charge on the grid is reduced to a low enough point, a few electrons will find their way through the wires

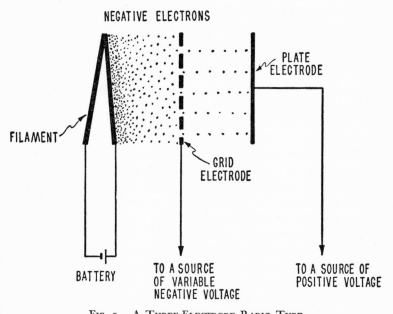

NEGATIVE ELECTRONS

PLATE ELECTRODE

FILAMENT

GRID ELECTRODE

BATTERY

TO A SOURCE OF VARIABLE NEGATIVE VOLTAGE

TO A SOURCE OF POSITIVE VOLTAGE

FIG. 9. A THREE-ELECTRODE RADIO TUBE

The grid electrode acts as a valve, regulating the number of electrons that are allowed to reach the plate.

of the grid to the plate. Thus a small electric current will flow in the plate circuit as it is called. A further reduction of the negative charge on the grid results in a corresponding increase in plate current flow. In practical radio tubes, it is possible to control a large flow of plate current with a small variation of charge on the grid. Weak signals received in a radio antenna can therefore be amplified a million or more times to provide strong signals to operate the loudspeaker.

69

Why do small animals require proportionately large quantities of food?

The amount of heat lost from a square inch of skin area is about the same for all warm-blooded animals. It is necessary, therefore, that animals have a food supply which is proportional to their surface, not to their weight. If one were to multiply each dimension of a mouse by ten, its volume (or weight) would increase one-thousand-fold while its surface area would increase only one hundred times. The larger mouse would require only one hundred times as much food as the smaller one despite the greater disparity in weight. A man weighs about as much as five thousand mice yet his average food consumption is only one-seventeenth that of the combined food eaten by the five thousand mice. This is because the skin area of a man is one-seventeenth that of the five thousand mice.

What does Einstein's theory of relativity mean?

It would be presumptuous to attempt to explain such a profound theory in the space available, but it is interesting to discuss some of its implications. The fact that all motion is relative has been recognized for a long time. A person traveling on a railroad train may consider himself in motion with respect to the rapidly changing countryside and at rest with respect to the gentleman sitting across the aisle. The existence of motion has meaning, therefore, only when considered relative to an object which may be thought of as fixed. Professor Einstein extended this fundamental concept and formulated what is known as the theory of relativity.

The theory states that (1) the motion of a body traveling at uniform speed through space cannot be detected by observations made on that body alone, and (2) the speed of light in free space is independent of the relative motion of observer and light source. An amazing consequence of the theory is that it implies (to physicists) an increase in the mass of an object as its speed increases. This increase in mass is insignificant, however, except when the body is moving very fast. If it were possible to accelerate an electron weighing .000,000,000,000,000,000,000,000,000,911 grams to the speed of light, it would weigh more than all of the planets, the sun and stars put together!

The theory also predicts that mass and energy can be transformed into each other. If one were to annihilate one ounce of matter (any kind will do) and convert it completely to energy, he would have made up 700,000,000 kilowatt-hours of energy; enough to supply all of the electric power to New York City for several weeks.

Where do tapeworms come from?
The tapeworm is a parasite that has neither mouth nor stomach. It lives in the intestine of its host where it "soaks up" food through its skin. Since this food is already digested, the worm has an exceedingly easy life. What is ordinarily regarded as a single worm is actually a colony of individuals connected end to end. Each section consists of an almost complete set of organs, with particular emphasis on the reproductive organs. Some kinds are only an inch or so long while others measure up to sixty feet. One of the longest kinds lives in the intestines of North American bears. It is also found in human beings who like to eat raw, pickled or poorly cooked fish. Even smoked fish may contain cysts of these tapeworms which develop after being eaten. Still another tapeworm is common in dogs. If this tapeworm gets into the human body, it produces large and dangerous cysts in the lungs.

Another parasitic flatworm is the beef tapeworm which seems to enjoy passing one stage of its life cycle in the intestine of man. As mature eggs develop in the worm, sections break off and pass out of the human body in the feces. If a cow should happen to eat grass contaminated with this material, the worm embryos develop and form "bladders" in the muscles of the cow. When these bladders are eaten by man, they hatch to produce a new tapeworm.

Control of the beef tapeworm can be affected by always cooking beef thoroughly, by inspecting and condemning meat that has bladders in it, and by sanitary disposal of sewage. The only sure way to avoid beef tapeworm, of course, is by refusing to eat rare meat. If this presents too great a sacrifice to us, we must depend upon the quality of local meat inspection and sewage disposal.

Why does the inside of your windshield sweat in winter?
At one time or another, you have surely noticed that water seems to collect on cold surfaces. Take the outside of a glass of iced tea, for example; or the inside surface of automobile windows. These phenomena all have their basis in the fact that our atmosphere, as a result of evaporation, always contains some moisture. This moisture

is not in the form of tiny droplets of water, but is an invisible vapor as truly gaseous as the air with which it mixes. There is a limit, however, to the amount of water vapor that air can absorb, and this amount varies with temperature. The higher the temperature, the greater the amount of water vapor that air will accept. If we take a certain amount of air at room temperature and saturate it with water vapor, our sample would have a relative humidity of 100 per cent. This means that it cannot accept any additional water vapor *at that temperature.* If we heat the sample, its relative humidity goes down and it is willing to absorb additional vapor. But what happens if we cool our saturated sample to a lower temperature? Since it was saturated to start with, the excess moisture must be converted back into water. This condensation occurs at the surface of a cooling object, such as the outside of a glass of iced tea. The temperature at which it begins to take place is called the dew point.

If several people are riding in a closed automobile on a cold morning, the warm air exhaled from their lungs contains a large amount of water vapor. As this warm, humid air strikes the cold windows, its temperature is reduced below the dew point and the excess water vapor condenses into water. It is this same effect which gives us the morning dew, nature's thoughtful way of providing her plants and flowers with their morning refreshment.

What are "atomic tracers?"
An atom consists of a nucleus around which whirl a number of orbit electrons. If the atom is stable, its nucleus contains the correct number of protons (just equal to the number of orbit electrons) and a quantity of neutrons which for some reason helps to make the nucleus stable. For a given element, there may be several quantities of neutrons that will provide this stability. Thus, for copper, with 29 protons, a stable nucleus can exist either with 34 or 36 neutrons. The two isotopes or "twins" so produced are known as Cu-63 and Cu-65, from the total number of nuclear particles, and both kinds exhibit the same chemical properties. Isotopes such as these are known as stable isotopes because their nuclei exhibit no erratic or unexpected behavior. But if we were to produce an atom with a nucleus containing the "wrong" number of neutrons, the nucleus would be unstable and would not last indefinitely.

So far we have differentiated between stable and unstable nuclei with respect to the length of time each would last. Just what does this mean? What happens to a nucleus when it decides to stop 'lasting." If the nucleus contains more neutrons than nature will allow, one of the extra neutrons suddenly changes into a (positive) proton. To accomplish this feat, it must somehow acquire a positive charge which is equal and opposite to the charge on an electron. It does this by manufacturing a negatively charged electron and throwing it away. Since the absence of a negative charge is the same thing as the presence of a positive charge, the neutron changes into a proton.

One of the best examples with which to illustrate the value of such radioactive materials is the man-made isotope of sodium, Na-24. This nucleus contains 11 protons and 13 neutrons, just one neutron too many for stability. It becomes stable by changing one neutron into a proton, ejecting a high-speed negative electron in the process. After the change has taken place, it settles down as a magnesium nucleus with 12 protons and 12 neutrons. In a given quantity of Na-24, it is impossible to predict when any one specific nucleus will eject an electron, but on the average one out of every two nuclei reacts in about fifteen hours. This means that the half-life period of Na-24 is about fifteen hours.

You will recall that sodium chloride, or common table salt, is made of sodium and chlorine. If some sodium chloride is made with Na-24 instead of the ordinary kind, each molecule is provided with an electronic label that can be read with a Geiger counter. This instrument indicates to the operator exactly how many electrons are being ejected at a given place, at a given time. Sodium of this kind becomes an "atomic tracer" which is proving invaluable to the physiologist. If a person is given a tiny amount of radioactive salt, that salt can be followed in its path through the body by means of a Geiger counter. In this way, investigators have determined the rate of blood circulation through the body, both in normal people and in patients requiring medical help. There are literally thousands of such applications, in many fields of science, where unstable isotopes can be used in place of their stable twins to help study scientific phenomena. Sometimes the job can be done in no other way. Sometimes atomic tracers are used because they provide the most economical and quickest solution to the problem. In any event, the nucleus is providing man with a new and effective ally in the quest for knowledge.

What is the history of the diamond?
References to the diamond have been found which date back to Biblical days. The Romans called them "adamas," which apparently progressed through "adamant," "diamaunt," "diamant," finally to "diamond." Ancient uses of the diamond included conversion of base metals to gold, curing of insanity and causing peace between husband and wife. It is still regarded favorably only for the last of these purposes.

The two main diamond fields of the world are in Brazil and South Africa. The Brazilian fields were discovered in 1727 and furnished most of the world's supply until the larger African fields were found in 1867. Scientists believe that diamonds may have been brought to the surface during the eruptions of volcanoes. It is also probable that they were formed under extreme pressure and heat during the formation of certain natural rocks. In 1796, Tennant and Wollaston in England proved that the diamond is composed of pure carbon by burning one at 850° C., collecting the gas that formed, and proving that the gas was carbon dioxide. Needless to say, this method has never been used extensively in the production of carbon dioxide for use in soft drinks.

Although the value of diamonds is controlled at a reasonably high level (about $1,000 for a one-carat stone), it has become the symbol of impending matrimony in this country, and as such will probably continue to be in great demand for many years to come.

Why does alcohol feel cool to the touch?
The temperature of a liquid, such as alcohol or water, is a result of the kinetic energy, or motive power, of its molecules. The greater the kinetic energy (and hence the motion), the higher the temperature. If, for some reason, a molecule having a particularly high amount of energy should pop out through the surface and escape, the average energy of the remaining molecules would be lower and the temperature of the liquid would go down. This is precisely what happens when a liquid evaporates. Alcohol feels cooler to the

touch than water because it evaporates at a much faster rate at body temperature.

How does a magnet work?

The nature of magnetism is suggested by an experiment which involves breaking a magnet in two, then breaking one of these parts in two, and so on. Tests of this sort have been carried out until only extremely small parts are left and these parts are always magnets in themselves. This suggests that if the process could be

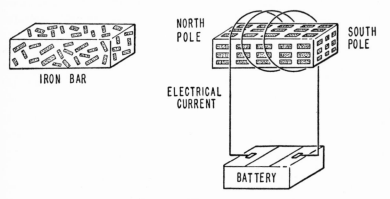

FIG. 10. MAGNETISM

On the left is a bar of unmagnetized iron, showing the random distribution of molecular magnets. On the right, a current of electricity produces a magnet by inducing all of the molecular magnets to line up in one direction.

carried out until parts of molecular size were obtained, these would presumably be magnets too. The theory that magnets are made up of molecular magnets is known as the molecular theory of magnetism.

According to this theory, an unmagnetized substance is one in which these molecular magnets are pointing in all conceivable directions and, consequently, exert no outside magnetism. To make a magnet from such a substance, it is necessary to align these little magnets so that their north and south poles all point in the same direction. This can be done by placing a bar of iron inside a coil of wire through which a current of electricity is flowing. The magnetic field produced by the moving electrons in the wire exerts

75

a force upon the molecular magnets causing all of them to line up in one direction.

A more complete theory tries to explain why some substances are magnetic while others are not. It is based on the fact that the atom consists of a heavy nucleus about which are revolving several lighter electrons. These electrons are believed to spin about an axis in addition to their rotation about the nucleus. This is analogous to the rotation of the earth on its axis as it follows its orbit about the sun. It is believed that magnetic substances, such as iron and nickel, have more electrons spinning in one direction than in the other. It is well known that moving electrons constitute an electric current and such a current produces a magnetic effect. If a given substance has ten electrons spinning in one direction and eight in the opposite direction, a net magnetic effect would result from the two unneutralized electrons. The application of an external magnetic field, as mentioned above, would align these molecular magnets so that their axis of magnetization would all lie in the same direction.

All of these concepts, however, are merely theories, or attempts to explain an observed condition. No one really knows what a magnet is or where magnetism comes from. It is truly remarkable that science has made such extensive use of magnetism while knowing so very little about what it really is.

Why are insects able to jump to such great heights?
Actually, the height to which an animal or insect can jump is almost independent of its size! Some insects, such as the grasshopper, can jump as high as two feet, a man about five. You will admit that this is a rather small difference in height when we consider the great disparity in size involved. As a matter of fact, insect muscles seem to be less efficient than those of man, or a grasshopper would be able to jump five or six feet in the air!

To explain this, let's go back to the grasshopper and analyze his ability to jump. This ability depends on the amount of jumping muscles possessed and upon the weight of the grasshopper. Measurements tell us that the average grasshopper has enough jumping muscles to push the weight of his body about two feet into the air. If we double his weight and at the same time double the weight of his jumping muscles, he would still be able to jump about the ame distance. Although his muscles produce twice as much force, they must propel twice as much weight into the air. If we were to continue this process of doubling his weight and muscle size long enough, we would obtain a man-sized grasshopper, but he would still be able to jump only about two feet off the earth. Although our imaginary grasshopper has about as much muscle for his weight as a man, he can't jump as well because his muscles are not as efficient as those of a man. Strange as it may seem, man is one of the best jumpers in the animal kingdom!

What happens to the carbon dioxide that people are continually exhaling?

The oxygen in the air we breathe combines with carbon within our bodies to form carbon dioxide. The uselessness of this compound to animal life is emphasized by the fact that animals have been exhaling it for hundreds of millions of years for want of a better use for it. And yet, chemical analysis of our atmosphere indicates that the carbon dioxide content is only about three parts in ten thousand by volume. Where does all of this carbon dioxide go? The answer lies in the life process of the plant world. Green leaves are the great sugar factories of the world. Only leaves containing chlorophyll are able to produce a certain kind of sugar called glucose. In order to produce glucose, the leaf must obtain the necessary raw materials for its manufacture. It has been determined that the formula for glucose is $C_6H_{12}O_6$, which means 6 atoms of carbon, 12 atoms of hydrogen and 6 atoms of oxygen. All of these elements are readily available to the plant in the form of carbon dioxide (CO_2) and water (H_2O). Through the process of photosynthesis, green plants are able to change these ingredients into glucose. Oxygen is a by-product of this process and is released to the atmosphere. Animals exhale carbon dioxide which plants need and plants give off oxygen which animals need. This cycle shows that the substances of life can be used over and over again by plants and animals, emphasizing the interdependence of the plant and animal kingdoms.

Who invented the microscope?

It is believed that the first microscope was made at some date between 1590 and 1607 by one of three spectacle makers of Middleburg, Holland—Hans Janssen, his son Zacharias, or Hans Lipper-shey. The man who first made extensive use of the microscope, however, was Anton van Leeuwenhoek who did his work over 270 years ago amid the windmills and canals of Holland. By a fortunate accident he came into possession of a strange instrument consisting of a long tube with a lens at each end—one of the early microscopes. With this instrument, and many better ones that he constructed himself, he made discoveries that opened up a new world of limitless and undreamed-of mysteries. It was this almost forgotten man who discovered the fantastic world of "wretched little beasties" living and dying in a world all their own. Anton Leeuwenhoek, a janitor and amateur scientist, was the first to see this invisible and completely fantastic world that had been completely hidden from the eyes of men from the beginning of time. His insatiable curiosity led him to look for his amazing microscopic creatures in the most unlikely places—in his mouth, in rain water, and even in shellfish taken out of the canals of Delft. He found life almost everywhere—in dust, in the earth and in the sea. Everything seemed to be teeming with microscopic life. Under the gaze of the microscope, even blood changed from a thick red fluid to a yellow liquid filled with floating, coin-shaped disks. Similarly, flesh, insects and plants, all changed from molded masses of "substance" to marvels of intricate beauty. His neighbors and friends thought that he was out of his mind, but his work paved the way for other great men of science. Although he failed to connect these micro-organisms with disease, he did provide others to come with the first great tool of their profession, a useful form of the microscope.

Where do calories come from?
One of the earliest studies of nutrition was made by Sanctorius in 1614. He weighed himself before and after meals to determine how much of the weight of his food had disappeared in the form of perspiration and heat. Investigators later discovered that food combines with oxygen in the body thereby releasing energy and that this energy could be measured. The unit of measurement adopted is the calorie, which is the amount of heat required to raise the temperature of one kilogram of water one degree centigrade.

The basic source of calories is glucose, a sugar produced by plants. Through a process called photosynthesis, carbon dioxide and water are made into glucose in the living cells of plants. The "machines" that manufacture glucose are the green plant cells containing chlorophyll. Like any other machines, however, they require energy to make them run. This energy is supplied by the sun. When the sun goes down, the manufacture of glucose stops just as a lathe stops when the electric power is turned off.

The process of glucose manufacture is vital to all living things, for without glucose neither plant nor animal could live. Of course, man has found other sources of calories such as fats and proteins, but these forms always find their way back ultimately to the first link in the food chain, green plants.

Why does the earth have seasons?

We might at first suspect that the variation in distance, throughout the year, between the earth and sun causes the seasons. A little reflection, however, will show that this cannot be the answer. On January 2 the earth is about three million miles closer to the sun than on July 2 but this should make the earth warmer if it had any noticeable effect at all. Although the Southern Hemisphere does have summer weather on January 2, the Northern Hemisphere is in the middle of winter.

The true cause can be understood with the aid of an analogy. If we hold a flashlight several inches from a wall and pointed directly at the wall, we notice a circle filled with light. If the flashlight is tilted slightly, however, the circle changes to an ellipse and covers a larger area. Since the amount of light coming from the flashlight has to be the same, the intensity of light on the wall must go down. This is because the same amount of light has to cover a larger area. An effect similar to this is the cause of our seasons. The earth rotates about an axis as it revolves about the sun and this axis is tilted with respect to the plane of its orbit. This means that the portion of the earth which is tilted toward the sun receives its warmest weather. Conversely, the opposite hemisphere will receive slanted rays and will, therefore, have its coldest weather. If the earth's axis were perpendicular to the plane of its orbit, there would be no seasons at all and the weather would progress from hottest at the Equator to coldest at the poles and the temperate zones would have springlike weather all the year round.

When were germs first associated with disease?
An unwillingness to associate microorganisms with disease existed for almost two hundred years following their discovery. As late as 1843, Dr. Oliver Wendell Holmes was ridiculed for his essay suggesting that "childbed fever" was caused by an invisible "something" carried from patient to patient by the doctor. It was not until the later part of the nineteenth century that the microbe received universal recognition for its role in the nature and cause of disease.

Where did the word "vitamin" come from?
Long before scientists learned what vitamins were, they knew that there was more to food than carbohydrates, proteins and fat. When they fed these substances to animals in the pure form, the animals lost weight and eventually died. Yet they thrived on milk which had no other known food value. Scientists theorized that there must be a single substance in milk which was vital to life. Not knowing what that substance was, they suspected that it was connected with another substance in milk, known as amine. Taking the *vit* from "vital," and adding *amine,* they called the unknown substance *vitamine.* Later, when the true nature of vitamins was discovered, they dropped the *e* and spelled it "vitamin." Today we know that there are many vitamins, not merely one. Lacking names for them, scientists identified them by letters of the alphabet. Still later, they began to identify the actual substances that make up vitamins and many of these are available on the market today. Although good food and sunlight insure an adequate supply of vitamins, it is comforting to know that our laboratories have made these vital substances available if we need them.

What is energy?
Energy, in a scientific sense, is anything that can be converted into work. Heat energy can be used to convert water into steam to run a locomotive. Electrical energy can be used to run a powerful motor, and atomic energy can be used to drive a battleship. All of these are examples of energy putting matter to work. Not all energy, however, is connected with motion. A piece of coal, or a stick of dynamite, certainly looks innocuous but each contains a considerable amount of energy. Energy of this kind which lies

dormant is called *potential* energy. The energy of bodies in motion is called *kinetic* energy.

The classification of energy as potential or kinetic is based upon whether the energy is stored up or whether it is due to the motion of a body. It is convenient, however, to further classify it by types, or forms. The most common of these are mechanical, heat, light, sound, chemical, electrical and nuclear. Energy is readily converted from one form to another in everyday life. As a matter of fact, everything we do is somehow connected with this conversion, even to the blinking of an eye. In our body processes, chemical energy from the sugars, starch, fats and proteins that we eat is burned to produce heat and the mechanical energy we use. Everything done in the world relies upon energy and its conversion from one form to another.

How can we determine the age of a fish?

The scales of a fish grow out from its body like the shingles on a house. They are set at an angle so that they overlap to form a complete scaly covering for the body. We can tell the age of most fish merely by counting the rings on the scales with the aid of a magnifying glass. This is possible because the scale's growth is seasonal, resulting in periods of slow growth followed by periods of greater growth. This produces variations in texture which correspond with the number of years the fish has lived. Don't try this method on your pet goldfish, however, since its scales should never be touched, let alone pulled out. A protective material covers the scales of live fish and if it is removed by improper handling, the fish may die. The next time you're having fish for dinner, why not take a look at the scales through a magnifying glass and proudly announce to the rest of the family the age of the main course that evening!

Where do soap bubbles get their color?

The beautiful pattern of colors seen on a soap bubble or on a spot of oil on a wet pavement is produced by a phenomenon known as the *interference* of light. This interference results from the fact that light is reflected to your eyes from two surfaces that are extremely close together. When you look at a soap or oil film, part of the light comes to you from the front surface while some has been reflected from the rear surface. If the two reflected rays come back to you out of step with each other, they cancel and no light reaches you from that point on the film. Whether or not this cancellation occurs depends on the thickness of the film. Why then, you might ask, do we see colors? Why not just dark and light areas due to variations in the film's thickness? If white light consisted of only one wave length, that is just what would happen. But since white light is really composed of colored light of many different wave lengths, the film is only able to cancel one of these colors at any given point. Since white light consists of many different wave lengths and since a soap film is not of uniform thickness, the result is that we see a variety of colors, each one being white light *minus* the particular wave length that is removed by interference. When green is removed, for example, the film will have a purple hue.

Can spiders fly?

Spiders engage in some of the most unusual occupations imaginable. The *water spider,* for example, breathes air like any other spider, but has mastered the art of living comfortably under water! He accomplishes this feat by building an inverted silken bag and attaching it to a limb of an underwater plant. Every time he makes a trip to his diving-bell home, he carries along many bubbles of air which cling to his body. Soon the underwater home is filled with air for the spider to breathe.

The *trapdoor spider* is perhaps even more ingenious. He digs cylindrical holes in the ground, lines them with silken walls, and seals the opening with silken trapdoors. The doors are sometimes hinged and equipped with internal grooves or "handles." When an enemy happens along, the spider merely holds the door shut by grasping the handle, thereby keeping intruders out of his spider-land home.

In addition to having mastered the art of living in water as well as on land, spiders have also taken successfully to the air. The *ballooning spider* has learned how to take flight and land almost at will. This would not be unusual if he had wings, but the fact is he doesn't. He spins extremely light silken threads which are lifted up into the air by the slightest breeze. The spider holds on and is borne along from place to place by this silken web. When he decides to land, the spider merely spins an additional thread which enables him to descend gradually, like a man being lowered out of a helicopter.

How do our ears determine the direction of a sound?

We are able to determine the direction of a sound source because we have two ears instead of one. When our head is facing the source of a sound, the sound arrives at both ears at the same time. This results from the fact that sound waves travel the same distance to either ear. If we are facing in a different direction, the sound arrives at one ear slightly ahead of the other because of the different distances involved. Our brain then uses past experience to translate the time difference into direction. To demonstrate this principle, we can roll up some newspaper into two tubes about five and six inches long respectively and two inches in diameter. While an aide produces a high-pitched tone (such as a whistle or the tinkle of a glass), place one tube over each ear. By exchanging the tubes from ear to ear we can make the sound appear to come from first one, then the other direction.

What is galvanized iron?

Iron that has been protected by a thin layer of zinc is called galvanized iron. Such items as trash cans, washtubs and water buckets would rust rapidly if made of unprotected iron. The zinc layer prevents moisture from coming in contact with the iron and preserves it indefinitely.

Paradoxically, zinc itself is very susceptible to corrosion. Fortunately, the zinc oxide which is produced on the surface soon forms a thin, white and quite impervious coating which protects the zinc from further corrosion. Since zinc compounds are somewhat poisonous, utensils made of this material should never be used for cooking purposes.

How does the ocean manufacture rocks?

Every river that flows into the ocean carries with it fragments of mineral matter, called sediment, which eventually separate and find their way to the ocean floor. The first sediments to settle out are the heavy gravels, followed by the sands and finally the fine silts and clays. While the process of assortment is never exact, the resulting deposits are fairly definite, with the heavier materials located closest to shore. When the deposits have accumulated to a depth of several hundred feet, the lower layers change into *sedimentary rock.* Two processes seem to be responsible for this change in form. Fine materials, like clay and silt, stick together under the tremendous pressure to form *shale.* The coarser materials, like pebbles and sands, do not stick together unless they are cemented. For this purpose, the ocean supplies dissolved minerals such as quartz, lime and limonite which are natural cements. These materials permeate the sediment, eventually bonding the fragments into one solid rock. As the sediments harden into sedimentary rock, the gravels change into *conglomerate,* the sands form *sandstone,* and the clays and silts form *shale.*

Another kind of ocean rock has its origin in the skeletons of living things. In deeper and clearer water, slightly beyond the limit of shale formation, great numbers of shellfish live and die. Over great periods of time, their shells accumulate at the bottom just as other sediments do. When the pressure is great enough, these fossil remains change into *limestone.*

How did Pasteur develop his treatment for "mad dog" bites?
Hydrophobia, or rabies, was an incurable disease at the time of Pasteur. He decided to search for a cure and began by injecting the hydrophobia virus into rabbits. He found that a greatly weakened, or attenuated, virus could be obtained by removing the spinal cords of inoculated rabbits and slowly drying the tissues. When this attenuated virus was injected into dogs, they became immune to the disease. The big problem, of course, was to determine if human beings could be made immune in the same way. Before he had time to experiment with some of the higher animals, he was presented with a dilemma. On July 6, 1885, a young boy who had been bitten by a mad dog was brought to him for treatment. To do so would be to expose a human life to an unknown and dangerous treatment; to refuse would be to sign the boy's death warrant. He finally decided in favor of the one chance the boy had for life. He administered fourteen injections of his weakened virus and waited for the results. Not a sign of the disease appeared.

What are voltage and current?
An electric current is a flow of electrons along a wire or other suitable conductor. It is analogous to the flow of water through the pipes in our homes. When electrons flow through the wire in an incan-

descent lamp bulb, they heat the wire, resulting in the emission of light. In many other ways, electric currents are able to do work for us and make our lives easier and more enjoyable. But why do they perform this work? An ordinary piece of wire contains many free electrons, but they refuse to move in any but random directions and, hence, will do no work for us. The situation is similar to turning off the main water valve in your home. Although the pipes are full of water, there is no pressure to force it out of the faucet. In order to induce electrons to do work, we must supply them with electrical pressure which forces them to move through the wires in the form of an electric current. This electrical pressure is called *voltage*. It is produced by an electric generator, or a battery which provides a continual force tending to push electrons through our appliances. The standard voltage supplied to American homes is approximately 117 volts. It is wired throughout our homes and made available whenever we want to use it. The amount of current we use depends upon the number and types of appliances that we use, but the voltage is always about the same.

Why are humans able to talk?

The human voice is perhaps the most amazing musical instrument of all, in spite of its simplicity. We are able to voice sounds because of two vocal chords located in the air passage between the mouth and the lungs. To understand how they work, imagine that we have a spool with two thin rubber strips fastened across one end so that their edges almost touch over the hole. If we blow through the open hole, a sound will be produced in much the same manner as one made by the vocal chords.

Ordinarily, the vocal chords lie open and out of the way of air moving in and out of the lungs. When we desire to speak, they become tightened and move together forming a slit through which the air must pass. This movement of air across the tightened slit produces a sound. The tighter the slit, the higher pitched will be the sound.

Sounds are changed into speech by the position and movement of the tongue, teeth and lips. Try saying the vowel "i" and you will notice how they move to make the sound. Some consonants, such as "f" and "s," are made without using the vocal chords at all. They are produced by exhaling air while the various parts of the mouth

assume appropriate positions. Speech is the combination of all these sounds into recognizable patterns which have arbitrary but accepted meanings.

What ingredients are needed to make an H-bomb?

Scientists have long known that the sun "makes" its energy by changing the lighter elements into heavier ones. To understand this principle, let's look at the two lightest elements, hydrogen and helium. A typical hydrogen atom has a relative weight of 1.008 atomic weight units, while helium, the next heavy element, has a weight of 4.003. If we were to change four hydrogen atoms into one helium atom, we would have 0.029 units of weight left over. This "left-over" portion of hydrogen would be converted into energy in accordance with Einstein's theory of the equivalence of matter and energy. When many such mass-to-energy conversions take place, we have an H-bomb.

Until 1945, when the first atomic bomb was exploded, man did not have at his disposal the means with which to achieve the tremendously high temperature required to make the hydrogen-to-helium change take place. It is estimated that a temperature of 20,000,000° C., comparable to that of the sun, is required for this reaction. With the creation of the atomic bomb, such temperatures became possible, and the hydrogen bomb was well within reach. While the details of the bomb are closely guarded secrets, the principles are well understood. They involve a source of high temperature, such as an A-bomb, and a supply of light elements which can be changed readily into heavier ones. This explains why the H-bomb followed so closely on the heels of its predecessor, the A-bomb.

Why are lenses able to bend light?

Many of the visual oddities of everyday life depend on the fact that light is bent, or *refracted,* from its course as it passes from one material to another. The pool of water that appears ahead of us on a dry concrete road, the apparent break in the oar at the surface of water, the famous mirages of the desert—all of these, and the very process of vision itself, are explained by the bending of light rays. Modern theory tells us that the refraction of light results from the fact that light travels at different speeds in different materials. To understand why this causes bending, imagine a column of soldiers, *side by side,* marching obliquely across a pavement. At the edge of the pavement, the land gets muddy and walking difficult. As the first soldier reaches the mud, he slows down. Then the one beside him reaches the mud and slows down, then the third, and so on. Since the soldiers on the pavement walk faster than the ones in mud, the column changes its direction, tending to pivot about the first soldier to reach the mud.

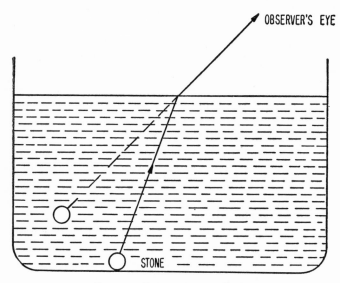

FIG. 11. REFRACTION OF LIGHT
Light coming from the stone is bent (refracted) at the surface, making the stone appear above and beyond its actual position.

The same thing happens to light rays as they reach the surface of a material requiring a change in the speed of light. If the new material has a slower speed, they pivot about the first ray to strike the surface. If the new material has a higher speed, they pivot about the last ray to strike the surface. In either case the direction of the light is changed, and we say it has been refracted.

Can sound be transmitted through a vacuum?
Unlike light, sound must have a transmitting medium in order to move from one place to another. Most of the time, sound comes to

us through the air. The process is analogous to a row of bent cards standing on end and arranged to fall in succession when the first is tipped over. If you remove several cards from the center of the train, the chain reaction will stop when the void is reached. In a similar manner, sounds produced in a vacuum will never reach your ears. A descriptive experiment involves placing a doorbell in a jar and then removing most of the air from the enclosure. When the bell is made to ring, no sound is heard because sound does not travel through a vacuum. As air is slowly allowed to enter the jar, the sound gets louder and louder until all of the air has been readmitted.

Although air is a conductor of sound, it is far from being the best. Most liquids are better conductors of sound than air. The sound of an outboard motor which is only barely audible through air is heard quite easily if you listen under water. Solids are even better, however, than liquids. Have you ever heard of Indians listening for hoofbeats with their ears to the ground? Trains also can be heard from a greater distance by listening to the sound coming through the rails. Among the gases, dense gases are better transmitters of sound than lighter ones and compressed gases are better than rarefied ones.

Do we learn best under stress?
Scientists tell us that stress has an adverse effect on our ability to learn. In an experiment, two groups of college students were selected who had the same ability to memorize nonsense syllables. These were three-letter syllables, such as "mog" and "lon," which have no meaning in the English language. Then a second test was employed. The experimenter tapped four wooden blocks with a fifth in a given pattern and asked the first group to reproduce the patterns. After they had done so, they were given more nonsense syllables to memorize. The same thing was done with the second group, except that its members were told they were doing very poorly with the block tapping, so poorly that they could not expect to get through college.

They began to show distress, and their performance with the cubes got worse and worse. When they were again tested with nonsense syllables, their performance had deteriorated very badly. They were now far inferior to the first group.

At this point, the poor group was given additional patterns to reproduce, and the experimenter encouraged and praised them for their good performance. Their ability rose, and when they were

88

again given syllables to memorize they had regained their former ability in full. There was now no significant difference between the two groups. We see from these experiments that stress debilitates our ability to learn and remember. We learn best in the presence of encouragement and praise.

How does the electric eye open and close doors?

When certain metals, such as potassium, sodium or cesium, are illuminated by rays of light, they eject electrons from the surface into the surrounding air. Imagine a plate of cesium which is illuminated and connected to the negative terminal of a battery. As rays of light strike the metal, many electrons pop out into the surrounding air. Now, let's attach another plate of an ordinary metal, like copper, to the positive terminal of the battery and bring the two plates close together without allowing them to touch. We now notice that the loose electrons are attracted to the positive (copper) plate. This movement constitutes a current of electricity which flows through the air from the negative to the positive terminal. The current of electricity that has been generated varies in magnitude with the intensity of the illumination. This *photoelectric effect* is the basis of "the electric eye," so widely used to open and close doors automatically.

The term "electric eye" was probably chosen because the light-sensitive material is always placed in an evacuated glass bulb to protect it from the atmosphere and to facilitate the required flow of electrons. Of course, the electric current so produced is hardly great enough to open or close doors. This is accomplished by various sorts of electronic equipment which magnify the current to a point where ordinary motors can be operated. In practice, a beam of light is directed across a doorway and picked up by an electric eye on the other side. The resulting current is amplified and used to prevent the door from opening. When a person's body interrupts the beam of light, the current drops to zero. This reduction in current is sensed by the associated equipment and a motor opens the door.

How far away are the stars?

Interstellar distances are so great that a special unit of measurement, the light-year, was invented to cope with them. A light-year is the distance that a ray of light will travel in a year. In terms of a familiar unit, a light-year amounts to the amazing distance of 6,000,000,000,-000 (six trillion) miles. The closest star, excluding our sun of course, is Proxima Centauri, which is about four and a third light-years or twenty-six trillion miles away. It is visible from the Southern Hemisphere. The closest star in the Northern Hemisphere is Sirius, eight light-years away. With the naked eye, we are able to see stars about 800,000 light-years away! Our powerful telescopes increase this distance over a thousand times, enabling us to observe objects located an incredible six billion trillion miles away. Light that we view from such distant stars left its source a billion years ago and has been traveling earthward ever since. When these rays of light were born, the only life on earth consisted of algae and other one-celled forms of marine life.

How are phonograph records made?
Although there are a variety of records on the market which play at different speeds, all are made by the same general method. Microphones in the recording studio convert the artist's performance into variations in an electric current. These electrical impulses are amplified and recorded on a master disk by means of an electromagnetic cutting head. This device consists of a sharp needle which cuts zigzag grooves in the master disk to correspond with the original sound variations. The master is then lacquered and a wax impression made of the grooves. The wax impression is dusted with finely powdered carbon (a conductor of electricity) and plated with copper by means of electricity. The original grooves cut on the master disk show up as ridges in the copperplated disk. That is, they are just the reverse of the original master. The copper disk is used as a mold to produce the final record. By pressing the mold into a plastic record material, an impression is made which is exactly like the original master. In this way, many records of good quality can be made at reasonable cost from one original master.

How are lagoons formed?
Lagoons are areas of quiet water between the mainland and an offshore bar. Such offshore bars, or barrier beaches, are found wherever straight shorelines of shallow water exist. They are the result of the action of wind and waves on shallow, sandy bottoms. Waves rolling in toward shore change into breakers as soon as the lowest

portion of the trough touches bottom. This scraping of the bottom picks up sand and deposits it a few feet closer to the shore. This forms a piled-up sand bar parallel with the line of breakers. As the action continues for many years, the bar grows larger and larger until it becomes a permanent feature of the shoreline. Long Beach, Jones Beach and Fire Island are well-known offshore bars along the south shore of Long Island, New York. The Great South Bay, which they make with the mainland, is really a lagoon.

Can snakes be charmed by music?

Since snakes have no ears, we can discount most of the snake charmer's claim that music induces snakes to dance. It's probable that the snake merely follows the rhythmic movements of the musician's body as he plays his instrument. In spite of their lack of ears, however, snakes do seem to respond to vibrations reaching them through the ground. Tests made with cobras serve to illustrate this point. With their eyes "blindfolded," the snakes turn and face in the direction of the experimenter's footsteps as he walks around them. In contrast to this response, the noise of a blaring bugle has no noticeable effect on their equanimity.

What gives the illusion of motion to a motion picture?

Some airlines have a colored stripe painted near the end of each propeller blade. When such a propeller spins rapidly, we see a complete circle of color. This is due to a characteristic of the eye called *duration of vision*. An image formed on the retina of the eye persists for about one-twentieth of a second after the object has been removed. The stripe on the propeller forms a circle because the image of the stripe in one position persists until the next blade moves into the same position. Motion pictures are possible because of the duration of vision of the human eye. In commercial systems, twenty-four pictures, or frames, are projected each second. Each succeeding frame is slightly different from the one preceding it. The illusion of motion results because our eye retains each image until the next one takes its place. Slow-motion pictures are taken at two or three times normal speed and then projected at the normal twenty-four frames per second. This spreads the motion out over a longer period of time. If it is desired to show a motion picture of a flower opening up, the individual frames are taken at a slow rate, perhaps one per minute, and then projected at normal speed. This speeds up the action to show in a short time, action which really took much longer to occur.

Is steel more elastic than rubber?
Although it may sound improbable, tests prove that steel is actually more elastic than rubber. Since this seems to violate one of our basic everyday concepts, perhaps we ought to start by discussing elasticity in general. If a ball of any material is dropped onto a hard surface (one which is unaffected by the blow), the ball will deform, at least momentarily. This deformation is caused by the energy of impact. If our ball is perfectly elastic, its molecules merely "borrow" this energy for an instant, and then return all of it to the ball by pushing it back into the air. Such a perfectly elastic ball rebounds to exactly the same height as that from which it was dropped. Lead is extremely inelastic, and would not rebound at all. The energy of the fall would all be used up in permanently changing the shape of the lead. It would finally end up as heat due to the friction between molecules of the lead. Elastic substances like steel and rubber, however, lose only a small part of the energy in frictional heating. If we compare these two materials for their ability to bounce from a hard surface, we find that steel has a considerable edge over rubber and hence is more elastic. If we were to repeat the experiment on a wooden floor, however, we would find that rubber bounces much higher. This is due to the nature of the floor itself. An ordinary wooden floor deforms under the heavy weight of the steel ball and absorbs the energy that would otherwise push the ball back into the air.

What is the difference between a sprain and a strain?
Both *sprains* and *strains* are really dislocations, sprains occurring in the ligaments and strains in the muscles. Sprains are usually accompanied by swelling about the joint and are difficult to distinguish from dislocations of the bone. They occur most often about the ankle or wrist. To relieve the pain and swelling, the injured part should be raised to reduce the flow of blood, and hot and cold cloths should be applied alternately for several hours. A tight bandage may be used to reduce movement of the joint, but it should be loosened frequently if the swelling increases. All sprains should have the attention of a physician at the earliest possible moment.

The treatment of strains consists of rubbing and relaxing the injured muscle. Massaging should be directed toward the trunk to stimulate the flow of blood. It should be light at first, but may be more vigorous after the pain has subsided somewhat.

What is a flame?

A flame consists of burning particles of a gas. The wax in a candle must be vaporized before it will burn. A piece of wood must be heated sufficiently to drive off combustible gases before it will burn, and gasoline is so combustible because it vaporizes rapidly. Materials that change easily into a gas are said to be *volatile*. Coal, coke and charcoal burn with little flame because they contain little volatile matter. They are nearly pure carbon, a substance that does not vaporize readily.

Most ordinary substances which burn contain carbon and hydrogen. Such fuels burn to produce carbon dioxide and water vapor. These products are exactly the same as those produced by our body in the utilization of food. The two processes differ principally in the speed and temperature at which they occur. The higher temperatures associated with burning produce a visual effect known as flame. This effect is due to the energy involved in the rapid disintegration of the fuel molecules into their component atoms, and their recombination into molecules of water and carbon dioxide.

Why doesn't an ice cube raise the water level in a glass as it melts?

Believe it or not, a melting ice cube will not raise the water level in your glass as it melts. If you don't believe it, put some water and an ice cube in a measuring cup before you read on. The reason for this oddity is found in Archimedes' principle of buoyancy, which states that a floating body will displace its own weight of water. If our ice cube weighs one gram, it will displace exactly one gram of water. Since the ice cube weighs one gram, the water formed in the melting process also weighs one gram. Since the water originally displaced by the ice cube weighed one gram, the melted ice water just fits into the now iceless volume. Unbelievable? Well, just consider that ice contracts when it melts and the solution will seem more reasonable. In this case, it contracts just enough to occupy the volume of displaced water. The water level, therefore, remains fixed during the melting process.

Is friction always undesirable?

We normally think of friction as the demon that wears things out and generally makes life difficult for us. To some degree, this is all too true. Friction is responsible for most of our automobile repair bills and countless other inconveniences that we must face every day. But is there any good side to friction? Can it be all bad?

Let's imagine that you are situated in the middle of a large sheet of frictionless ice. What can you do to get off? No matter how hard you push against the ice, nothing happens, since there is no way to "get hold of anything." No matter how you scramble in your attempt at freedom, you find yourself helpless because there is no frictional resistance for you to push against. You appear to be marooned where you are forever. Actually, it's not quite as bad as it seems because you can get off quite easily by merely blowing some air out of your lungs. Just as in the case of a rocket engine, you will be pushed quickly from the sheet of ice back into the wonderful world of friction.

It would be a strange world indeed if there were no friction. We could not walk, the fibers of our clothing would fall apart and we would find it difficult to build houses because all of the nails would pop out as fast as we hammered them in. Screws and bolts would not hold firm because there would be no friction to hold them together. Objects would slide off the table and dresser unless their tops were absolutely flat. Without doubt, if friction is an evil, it's a very necessary one. You can use that fact for consolation the next time you have your shoes repaired.

What is high fidelity sound reproduction?

We hear a sound because the air in contact with our eardrum has been made to vibrate. The rate at which this vibration takes place is called the frequency (number of vibrations per second) or pitch of the sound. The higher the frequency, the higher the pitch. The average individual can hear sounds ranging in frequency from about 30 to about 15,000 vibrations per second. Ordinary music consists of a combination of many frequencies produced by various musical instruments. In order to have a high fidelity system, we must be able to reproduce all of these frequencies from the lowest-pitched note of the violoncello to the high-frequency tinkle of the smallest chime. Although modern recordings have a top frequency of about 15,000 vibrations per second, ordinary record players are only able to reproduce frequencies up to about 6,000 vibrations per second. Reproducing equipment of this kind loses a great deal of the original content of the music. High fidelity, or hi-fi, equipment is capable of faithfully reproducing all of the frequencies contained on the record. In purchasing such equipment, however, it should be borne in mind that there are no rigid specifications connected with hi-fi and, consequently, the quality of commercially available systems will vary considerably. It is wise to obtain competent advice before embarking on a hi-fi program.

How many active volcanoes are there?

Scientists believe that there are in the neighborhood of five hundred active volcanoes at the present time. The most active appears to be Mount Etna, in Sicily, which has had eighty eruptions in recorded history. Some have been extremely destructive, killing as many as ten thousand people at a time. Geologists believe that Mount Etna has been this active for hundreds of thousands of years.

The cause of volcanoes is connected with the very high temperatures that exist within the depths of the earth. The intense heat melts rock to a half-liquid state known as *magma* (from the Greek meaning "dough"). Steam and subterranean gases build up tremendous pressures which force the magma toward the surface. Closer to the surface, additional water is changed to steam, adding to the forces pushing relentlessly toward an outlet. At last a weak spot is found in the surface and molten magma, boiling at 1000° C. or more, pours out onto the land. As the great pressures are relieved, the flow slows down and the material hardens to "cork" or plug the mouth of the volcano. Whether it is dead or merely dormant, only time can tell.

Why do chickens lay so many eggs?

Since our domestic chicken is a bird, we might start by asking: "How many eggs does a bird lay?" There is no simple answer to this question. Each species has a *normal* number that usually seems to satisfy, but variations still manage to creep in. Take the thrush, for example. Northern thrushes usually lay four eggs a season, while their relatives in warmer climates stop at two or three. If a bird's nest is robbed, the bird will often lay additional eggs to replace the stolen ones. There is a case on record of a woodpecker laying seventy-one eggs in seventy-three days on this account! In the case of our everyday chicken, egg-robbing has virtually changed it into an egg-laying machine. Some hens lay more than two hundred eggs a season if they are promptly taken away as they are laid. If, on the other hand, they are not removed immediately, the hen will have had enough of the game after fifteen or twenty. At that point she will stop laying and concentrate on incubating what she has. As for the number of eggs that other birds have at one time, we may say that a full set seems to vary from one to about twenty.

What insect gives us shellac?
The raw material for shellac comes from an insect that lives in some of the Oriental countries. It is called the *lac insect* and is found principally in India and the Netherland Indies. The tiny red larva of these insects grabs hold of the bark of young twigs to feed on the sap. In its life process it secretes a resinous substance and forms a cocoon-like shell which joins with the shells of other larva to form a hard coating, or scale, on the twig. A resinous material, called *lac,* is obtained by removing these incrustations which are melted, purified and solidified to form shellac. Lac is also used in the production of sealing wax and a certain crimson lac dye.

Other insects that contribute raw materials to industry are the gall wasps, gall flies and gall lice. These creatures lay their eggs in the tender leaves and stems of many plants ranging from the blackberry bush to the oak tree. The larvae that develop feed on the surrounding tissues of the host resulting in the formation of protective swellings known as galls. These galls are collected and used in the manufacture of pigments for ink.

Why don't embryos look more like the parents?
All animal life comes into existence in the form of a single cell. In that one cell are all of the chromosomes and genes which determine the nature and characteristics of the individual. In all of the higher forms of animal life, this cell develops into the embryonic stage, at which time it looks very little like its parents. Scientists have observed that in the earlier stages of development, all embryos seem to follow the same general pattern of growth. The embryos of the higher

animals resemble those of the lower animals, at times exhibiting unnecessary organs and structures which later disappear. As an example, mammal embryos sometimes show signs of slits in the neck which resemble the gill slits of a fish embryo. In fish, these develop into gills, but in mammals they eventually close up and disappear. This tendency for the embryos of higher animals to resemble those of the lower ones is called *recapitulation*. It is interesting to note that the embryos of higher animals never resemble the lower adult forms, but only the lower embryos.

Why do light objects look larger than dark objects?
When a ray of light strikes the light-sensitive elements within the eye, the effect of the light spreads out over a slightly larger than

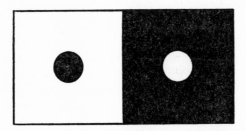

FIG. 12. THE APPARENT SIZE OF LIGHT AND DARK OBJECTS
The white circle looks larger than the dark one, even though both have the same diameter.

it ought to. It is as though the elements at the edge of the image reflect light to the side, causing additional light-sensitive elements to respond. For this reason, when we look at a light object on a dark background, the image "spills over" onto what should be a darkened portion of the retina. This makes the object look larger to us. In reverse fashion, a dark object on a light background looks smaller than it really is because light from the background encroaches upon the darkened portion of the retina. You can prove this to yourself by cutting identical small squares from pieces of white and dark paper. Then place the white spot on the dark background and vice versa. The light piece will look larger.

97

Why does a clinical thermometer "hold" its reading when removed from the source of heat?

Anyone who has used a mercury thermometer knows how elusive the silvery column of mercury can be. Have you ever seen a nurse remove a thermometer from a patient's mouth and search for the almost invisible column of mercury? Because of this loss of time, the reading of an ordinary thermometer would have changed enough to make the measurement worthless. This problem is solved by the use of a clinical thermometer which registers the highest temperature attained during a given period of time. This is accomplished by placing a constriction in the bore of the glass, a short distance above the bulb of mercury. Heat forces the mercury past the narrow constriction with relative ease, because of the large pressures produced by the expanding liquid. Going back down, however, is quite another thing. There is no force except the small weight of the liquid itself to push it back into the bulb. Once the thermometer reaches its highest reading, there is a full column of mercury from the high point down to the constriction. The column will remain there until the nurse snaps it back into the bulb by generating the required amount of centrifugal force.

Why does a baseball bat sometimes sting a player's hands?

Baseball bats, hammers, axes and, in fact, almost any object that is swung, are all forms of the *compound pendulum*. You will recall that a simple pendulum consists of a weight at the end of a "weightless" string. A compound pendulum is one in which the weight of the suspending element, or handle, cannot be neglected. If we hang a baseball bat from its handle end, and let it swing back and forth, we find that it swings somewhat faster than a simple pendulum of the same physical length. This results from the fact that a bat has its weight distributed throughout its length rather than having all of it at the end. Now let's shorten the simple pendulum, so that it swings at the same rate as the bat. If we measure this distance off on the bat from the point of suspension, we have located what physicists call the *center of percussion* of the bat. This is the point at which all of the bat's weight may be considered to be located for pendulum purposes. When the batter strikes a baseball at the bat's center of percussion, the knowing bleacherites comment that "he put good wood on it."

When the ball strikes the bat at some other point, the bat shudders and shakes, passing its indignation along to the batter's hands.

Another useful application of this principle can be found in the disciplinary use of the one-foot ruler. Although this may be a lost art, the teacher with a scientific bent will find that the most efficient point of application to the student's hand is the center of perscussion, two-thirds of the way from the point of suspension.

Is it harmful to drink water with meals?

The common belief that the drinking of water with meals is harmful has been largely discounted by scientists. Experiments have shown that digestive juices are just as effective in dilute solutions as in more concentrated ones. The one possible danger of drinking water with meals is that it may be substituted for adequate chewing. In such cases, food is not properly mixed with saliva, the substance which starts the digestion of starches and sugars. As long as food is chewed properly, water, when taken with meals, actually aids in the digestive process.

How do neon lamps work?

Everyone is familiar with the various sorts of "neon" lamps that are so popular today in advertising displays and signs. But where does the light come from? And why does it have its characteristic color? Lamps of this kind contain a gas through which flows a current of electricity. As the moving electrons, which constitute this current, move through the gas, they collide with gas atoms and impart some of their energy to the atoms. When this happens, some of the orbit electrons, which normally circulate around the nucleus of the atom, are dislodged from their customary positions. Atoms containing such disturbed orbit electrons are said to be in the *excited state*. After a short length of time, the excited atoms lose their excess energy and snap back into their normal positions. Each time such a return occurs, a bundle of light is produced and emitted. Strangely enough, the light produced by this process has a color which is characteristic of that particular element. Each gas has its own group of light colors and these may be used to identify the gas even in stars millions of light-years away. The balloon-lifting element helium was discovered in the sun by this method before its presence on earth was even suspected.

How is it possible for a jet fighter plane to accidentally shoot itself down?

The rapid advances in aircraft design have added a new hazard for jet fighter pilots: shooting themselves down. Amazing as it may seem, aeronautical engineers have made it possible for a supersonic jet fighter plane to "catch up" with the fire from its own guns with sufficient speed to shoot itself down. If a plane, flying at 1,000 mph fires a burst from its 20-millimeter guns, the shells leave the plane with an air speed (plane's speed plus muzzle velocity) of about 3,000 mph. They soon slow down, however, due to wind resistance, and the plane begins to overtake them. Normally, the shells drop sufficiently, because of gravity, to permit the plane to pass harmlessly over them. If the pilot fires his guns in level flight and then goes into a dive, however, it is possible for the plane to overtake and run into them with considerable force. If the shells happen to be explosive, great damage can be done to the plane. There is at least one instance known of a jet fighter pilot having shot himself down by this method.

What is the principle of the atomic reactor?

An atomic reactor, or pile, is a device which controls the rate of disintegration of uranium-235, and hence the rate at which energy is released. Thus, atomic fission is controlled to take place at a slow and desired pace. Each time an atom of the fissionable material is struck by a stray neutron, it breaks up to form several lighter elements, a great amount of energy, and a few "left-over" neutrons, the same kind of particle that caused the disintegration. The key to the operation of a reactor is the careful handling of these secondary neutrons.

In the nuclear reactor a *moderator,* such as pure graphite, is placed in such a position that most secondary neutrons must pass through it to reach additional uranium-235 atoms. Graphite used in this manner slows up the neutrons in their path toward new collisions. The control of the reactor is provided by rods of the metal cadmium which are inserted a variable distance into the pile. Cadmium has a great affinity for neutrons and, with the rods inserted full length, enough neutrons are captured to effectively eliminate the production of energy by secondary neutrons. As the rods are gradually withdrawn, more and more neutrons become available to produce fission of additional uranium-235 atoms. The energy production, therefore, increases as the rods are withdrawn.

Our present reactors produce heat energy which converts water to steam as in ordinary oil-fired boilers. Perhaps someday scientists will find an economical way to convert nuclear energy directly into elec-

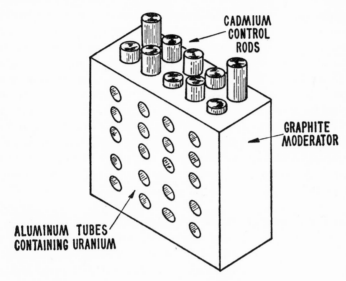

FIG. 13. A NUCLEAR REACTOR

Nuclear reactors use a uranium chain reaction in the production of heat. The moderator reduces the speed of the available neutrons, making them more effective in producing secondary disintegrations. The cadmium rods control the availability of these neutrons. When inserted into the reactor, they absorb neutrons, preventing them from producing secondary disintegrations. The rate of energy production can be controlled by varying the depth to which the cadmium rods are inserted into the reactor.

tricity. This problem is no doubt being considered today in many of our finest nuclear laboratories.

What are Cinerama and CinemaScope?

Cinerama produces an illusion of depth by making the projection screen very large and curved. This makes the audience feel a part of the picture. In addition, loudspeakers are placed at various points around the theater to aid in the illusion of depth. The picture is taken by a special camera which has three different lenses and three separate strips of film. The lenses are placed at an angle with one another in order to produce pictures of the left, center and right portions of the scene to be photographed. The completed films are projected by three synchronized projectors which reproduce the three portions of the scene on the screen.

CinemaScope also uses a wide-angle curved screen. It is the same height as an ordinary motion picture screen but twice as wide. A single projector, equipped with a special wide-angle lens, is used to form the image on the screen. Although the film used for Cinema-Scope is the same as for conventional motion pictures, the image produced has twice the normal width. This is accomplished by a rather clever technique. The lens of the camera which photographs the scene compresses the width of the picture while not affecting the height. This makes it possible to photograph a scene that is twice as wide as normal. When the picture is reproduced, the projector reverses the process by spreading out the width dimension of the image so that it appears in normal proportion. CinemaScope makes it possible to double the width of the picture with a minimum of obsolescence to existing motion picture equipment.

Why does a thin film of glue make a stronger joint than a thick one? When we succeed in getting molecules close enough together, we find that there is a considerable force of attraction between them. This molecular force of attraction is called *cohesion* if the molecules are of the same kind, and *adhesion* if they are not. The forces involved, however, are really versions of the same thing. If we place a plate of glass on the surface of water, it takes a considerable force to lift it up. This is a result of the adhesive forces between the glass and the surface of the water. Similarly, if we place two highly polished pieces of lead in close contact with one another, we find a force of attraction trying to keep them together. This is because of cohesive forces. In the case of glue, we must depend both on cohesion and adhesion. The glue must stick, or adhere, to the surfaces to be glued, and, of course, it must cohere to itself. Strangely, in most situations where gluing is indicated, the glue has stronger adhesive forces than cohesive forces. An excess of glue merely increases the chances that particles of the glue itself will pull apart.

How does a "gastroscope" look inside the stomach of a normal person?

When light enters an ordinary glass or clear plastic rod, it is reflected back and forth internally until it emerges from the other end. Such a "light pipe" can be bent into any convenient form with the same result. Dr. N. S. Kapany, of the University of Rochester, reasoned that this familiar principle could be used to transmit detailed images if many microscopic "light-pipes" were tied into a bundle. He conceived the idea of bunching a quarter of a million individual strands, each only a thousandth of an inch in diameter, into a device that could transmit pictures. Each strand is used to transmit a single point of light. The thousands of points of light combine to form an image in much the same way that ink dots on a newspaper form a picture. Using this principle he succeeded in producing a "gastroscope" which fits down the throat to give a close-up view of the human stomach. The device is capable of passing exact images from one end to the other even if it is tied in knots.

These techniques, which are referred to as *fiber optics,* can even be applied to code work. If a million fibers are woven *at random* into a bundle, the image emerging at the other end will be hopelessly mixed up. This is because each fiber has a different relative position at either end of the bundle. In order to unscramble the picture, it would be necessary to have another bundle exactly like the first with which to reverse the scrambling process. If two such tubes were connected by a television link, they would provide a virtually unbreakable code.

How were underground caverns formed?

Most of the famous caverns owe their existence to the eroding action of water on limestone. As water seeps into the earth from rain and streams, it must flow through cracks in the subsurface soil and rock. This water always contains a certain amount of carbon dioxide, the gas that animals exhale. Such a solution of carbon dioxide and water forms a weak acid, carbonic acid, which helps to dissolve the limestone. As the limestone dissolves, it is carried off by water and the cracks get larger and larger. Finally, after great lengths of time, large holes or caverns are formed.

Stalactites are an interesting by-product of this process. As a drop of water seeps through the rock and reaches the top of the cavern, it must either remain there and evaporate, or drop to the floor. If it evaporates from its position on the ceiling, a small amount of dissolved mineral matter remains. As drop after countless drop repeats this process, a stalactite, or mineral "icicle," is formed hanging from the ceiling. If the drop falls to the floor, a *stalagmite,* or icicle in reverse, is formed. It sometimes happens that a stalactite and stalagmite eventually meet to form a beautifully shaped column of stone reaching from floor to roof of the cave. The natural colors of the minerals involved add to the beauty of these formations.

Why do steel ships float?

People have always been aware of the buoyancy of water, but the discovery of its true nature was made by a Greek scientist, Archimedes, somewhere around the year 250 B.C. Archimedes found by his experiments that a body immersed in water is buoyed up by a force equal to the weight of water displaced by that body. Since a steel ship has many air spaces within its compartments, its average density is much lower than that of water. As it sinks lower and lower in the water, a point is reached where it displaces just enough water to balance its own weight. It then floats at that level.

If an object has an average density greater than that of water, like a rock or bar of iron, it will sink. Archimedes used this fact in the first practical application of his principle of buoyancy. The king had just had a new golden crown made and, for some reason, he suspected the goldsmith of mixing a less expensive metal with the gold. So he asked Archimedes to test the purity of the gold without damaging the crown. After puzzling over his problem for a while, legend tells us that the solution came to him while bathing and he ran out into the street naked shouting "Eureka!" ("I have found it!") Here's what Archimedes did. He weighed the crown (in air) and then borrowed an amount of pure gold just equal to the crown's weight. He then immersed the pure gold in water and noted the reduction in weight due to buoyancy. When he placed the crown in water, he found that the weight reduction was greater than for the pure gold. Evidently, the crown was made of a mixture of gold and some other less dense metal.

Are mineral waters an aid to good health?

In spite of the common belief that mineral waters are healthful, scientists tell us that their value is doubtful. Minerals that the body needs are usually well supplied in the diet, with the possible exception of calcium. Very few mineral waters, however, claim to supply calcium to their users.

Mineral water comes from the mineral springs that abound in many parts of the world. They vary from soda springs, which contain bicarbonate of soda, to sulfur springs, which contain hydrogen sulfide (also contained in bad eggs). The real value of mineral water probably lies in the health resorts usually connected with such springs. In general, the body needs a reasonable amount of exercise, plenty of rest, a good diet and a favorable state of mind. Health resorts are certainly capable of providing these needs. Where such resorts have effected cures, the "miracle" is probably attributable to the change from poor to good habits of healthful living. As for the water, drink it if you like it.

Why does wood burn?

Have you ever wondered why a piece of wood or a gas flame keeps burning away of its own accord? We know that such fuels are made of compounds containing hydrogen and carbon atoms which burn by uniting with oxygen atoms from the air. But why do they prefer oxygen to their present atomic associates? You will recall that most of the fuels we burn—coal, oil, paper, etc.—are all derived from living substances. Coal, for example, came from prehistoric vegetation. The hydrocarbons, so formed, are compounds of hydrogen and carbon which require the energy of sunlight for their formation. This energy is stored in the fuel and is released as heat when the hydrocarbons are changed back into simpler and more stable compounds like water vapor and carbon dioxide.

Fire is analogous to the roller coaster which, once pulled to the top of the hill, rushes down the slope with ever increasing speed. We must start a fire with a match or other source of heat. This tears the hydrocarbons apart into hydrogen and carbon which quickly unite with oxygen to form water and carbon dioxide. These products of fire have considerably less energy than the hydrocarbons that burned, so the excess energy is released as heat. This heat, in turn, causes further disintegration of the hydrocarbons and the process is self-perpetuating until all of the fuel is used up. Such a process is called a chain reaction since, once started, it will keep on going indefinitely.

What is "sonar"?

Sonar, which means *so*und *n*avigation *a*nd *r*anging, is a method which uses sound-wave echoes to determine the presence and location of submerged objects. Bursts of sound energy are transmitted by a source on the ship's hull. After a short period of time, echoes return which are picked up by the sonar receiving equipment. By measuring the time elapsed between sending out the sound and receiving it back again, it is possible for the equipment to determine the depth of water, or the presence of mines or submerged submarines. Sound travels at about 4,700 feet per second in water. If the time interval for the sound's round trip is two seconds, the object encountered must be 4,700 feet away. In this way, the sonar set is able to convert time intervals to distances for automatic presentation to the operator. The sound waves normally used in sonar systems are above the range of human hearing, in the *ultrasonic* range. In this respect they are somewhat like dog whistle "sounds" which, although audible to some animals, are completely inaudible to humans.

Who invented dynamite?

Dynamite is a mixture of nitroglycerin and a fine absorbent material which renders it less susceptible to shock. It was perfected in 1864 by Alfred Nobel, of Sweden, who is perhaps best known to us as the founder of the famous Nobel prizes. Nobel found that ordinary nitroglycerin could be handled much more safely if it was mixed with certain fine mineral substances. Such a mixture retained most of the explosive power of nitroglycerin while making its use much safer and, consequently, more widespread. Today, most dynamites use wood pulp as the absorbent material. Mineral matter is inert and can add nothing to the force of the explosion, while wood pulp burns to form gases which increase the amount of developed energy. In its familiar form, dynamite is molded into cylindrical sticks and covered with paraffin paper.

The basic explosive in dynamite, nitroglycerin, was invented in 1846 by Ascanio Sobrero, an Italian. It is an oily liquid made from glycerin and nitric acid. Please don't try to make it, however, as it is extremely unstable and the slightest shock results in the rapid production of great volumes of gases. It can be handled with safety only by experts.

What is the continental shelf?

The continental shelf is the relatively shallow region between the shore and an abrupt drop to the ocean floor. It can be thought of as an underwater extension of the land. Most of the continental shelves were once dry land, inundated by later advances of the sea. It is here that sediment is deposited by rivers to be turned into rock by time and pressure. And it was on these shelves that the great ice sheets of past ages gave up their burden of rock, soil and melted ice.

Our Pacific shelf is about 20 miles wide, while that of the Atlantic varies from about 150 miles in the north to a very narrow ledge in the south. The widest of them all measures 750 miles on the Barents Sea in the Arctic. The depth of the continental shelves starts at zero on the coast to perhaps a maximum of 1,500 feet. At the edge of the shelf we come to a sudden drop of between 2 to 5 miles to the ocean floor.

The contributions of the shelves include some of the world's best fishing grounds, tidewater oil fields, and great "farms" of seaweed used in the food, drug and fertilizer industries. This seaweed is so rich in mineral matter that scientists see in it a promising source of nutritious food for the future.

How can mosquitoes walk on the surface of water?

There is a natural inclination for us to distinguish between liquids and solids on the basis of their differences. But if we think of their similarities for a moment, we find that in both forms of matter there is a tendency for the molecules to stick together. While this force of attraction is much greater in solids, it does exist, nevertheless, in liquids. A molecule at the center of a liquid is attracted equally in all directions by its neighboring molecules. There is no unbalanced force acting on the molecule. Those at the surface, however, are attracted downward by their neighbors below without any compensating attraction from above. This provides a force on every surface molecule which tends to pull it from the surface, into the liquid. This means that the quantity of molecules on the surface is always at an absolute minimum. It is as though the whole surface were covered with a thin elastic skin tending to draw itself as tight as a drum, resisting any effort to increase its area. This characteristic of liquids is called *surface tension* and it is due to the cohesion of one liquid particle to another. When a mosquito steps onto such a surface, his weight tends to stretch the "film" against this surface tension and increase its area. Surface tension resists any increase in surface area by pushing up against the mosquito's feet. This keeps him safe so long as his weight is not great enough to break through the surface and rupture the film.

How can fear be controlled?

The word "controlled" was selected for the question instead of "eliminated" because some fear is important in promoting prudence and safe living. Psychologists advise us to deplore only the unrealistic, exaggerated fears—those involving highly improbable occurrences that constitute unnecessary obstacles to normal living.

A variety of methods have been tried to control such fears. Perhaps the most effective is *reconditioning*. This method makes use of the pleasant associations that normal people form with certain situations. Take Johnny, for example. At the age of four he has acquired a fondness for ice cream but he fears all small furry animals. To overcome these fears, Johnny might be given ice cream while a rabbit in a cage is brought as close as possible without disturbing him. Each day, as he eats the ice cream, the rabbit should be brought closer, until one day Johnny will find it possible to touch it or even hold it in his lap. By a slow process of reconditioning, the fear will eventually disappear. Strangely, related fears seem to disappear too. Once Johnny discovers that rabbits are fun, he also finds it easy to like squirrels and puppies. This case, of course, seems to illustrate that mental health is the result of continued and effective work, not merely isolated treatment. And of equal importance to the cure is the removal of those factors in his environment which made Johnny afraid of small animals in the first place.

How does a submarine submerge and come to the surface?

A submarine can submerge and surface at will because of the principle of buoyancy. By allowing just the right amount of water to enter its tanks, the submarine can make its total weight exactly the same as the water it displaces. This means that the submarine's overall density, or weight per unit volume, is the same as water. If the submarine is then placed at any given spot beneath the surface of the ocean, it will remain there indefinitely. It has no tendency to float, and no tendency to sink. If the engines are turned on, its forward velocity, together with the action of its rudders, will enable it to select any desired depth with ease. When it wishes to come to the surface again, part of the ballast water is forced out of the tanks by means of compressed air. Since the density of the submarine is reduced, its buoyancy enables it to rise to the surface. Because of the difference in density between sea water and fresh water, submarines must be extremely careful when navigating submerged near the mouths of rivers.

Why are some persons color-blind?

Color blindness is an eye condition found in some people which causes them to confuse two or more colors which others can readily distinguish. The defect occurs in about 6.5 per cent of men but in less than 1 per cent of women. It is usually inherited, although certain diseases and drugs are known to produce temporary color blindness. The most common form of the condition is the inability to distinguish between red, green and yellow, or between blue-green, blue and violet. According to the *Young-Helmholtz theory* of color vision, the eye contains three sets of nerves which respond to the primary colors of light, red, green and blue-violet. If all three sets are equally stimulated, we receive the sensation of white. If predominantly green light reaches our eye, it stimulates the green nerves to a greater extent than others and we receive the sensation of green. When yellow light reaches the eye, *both* the green and red sets of nerves are stimulated and we see yellow. The ability to distinguish between red, yellow and green, therefore, depends upon the presence in the eye of two sets of nerves, the red and green sets. If one of these is absent, or defective, we cannot distinguish between the three colors mentioned. In this way, all of the hues and shades are produced by the appropriate stimulation of one, two or three sets of color nerves. Rare forms of color blindness result from the lack of two or, in extreme cases, of three sets of color nerves. In such cases, a person may be completely color-blind—living in a world illuminated entirely by uninteresting shades of gray as in a conventional black and white moving picture.

Why do hydraulic brakes work so well?

The hydraulic brakes on automobiles are an application of *Pascal's principle,* which tells us that pressure applied to a confined liquid is transmitted undiminished and equally in every direction. The force of your foot on the brake pedal depresses a piston in the master cylinder. This produces pressure within the cylinder forcing oil through strong tubes to each of the four wheels. At each wheel, this pressure forces brake shoes against the wheel thereby producing frictional forces tending to stop the wheel's motion. When the foot pedal is released, the pressure in the hydraulic system is relieved and springs return the brake shoes to their normal position. Such a system is advantageous because equal pressures (Pascal's principle) are automatically produced at each wheel when the brake is used. It is easier to keep such a system in proper working order, and uneven braking is kept to a minimum.

What was the loudest noise ever heard?
The loudest noise ever heard, at least in modern times, was the eruption in 1883 of the volcanic island of Krakatoa, in the Dutch East Indies. It was much louder than any atomic or hydrogen blast to date. After being dormant for two hundred years, the volcano erupted with such violence that it was audible in Australia two thousand miles away. The violent explosions lasted for thirty-six hours and blew off half of the island. The final stupendous outburst on August 27, 1883 had enough sound energy to circle the earth completely, not once, but seven times before it faded out. It left a record of its passing on all of the self-recording barometers in the world. Cinders and ash shot twenty miles into the atmosphere making it dark for one hundred miles around, even at noon.

But even worse than the eruption were the tidal waves which followed. Hundreds of villages were wiped out by huge waves that rose one hundred feet in the air. The gigantic waves, rushing forward at speeds up to seven hundred miles per hour, finally dissipated themselves on the Australia and California coasts, thousands of miles away.

What is the difference between rheumatism and arthritis?
The word "rheumatism" comes from the Greek *rheum,* "a watery discharge." Rheumatism affects the soft tissue *outside* the joints—the ligaments, nerves, muscles, tendons and bursae, the fluid sacs between the joints. It is sometimes accompanied by a swelling due to the accumulation of water in the joints.

Arthritis, from the Greek *arthro,* "joint," and *itis,* "inflammation," is one of the oldest diseases known. Scientists have detected evidence of the disease in the bones of dinosaurs that lived over 200,000,000 years ago. Both the Java Ape Man and the Neanderthal Man suffered from arthritis. Museum mummies reveal signs of arthritic bone deformities in ancient Egyptians. As you probably have guessed, arthritis occurs *within* the joint that connects two bones. The disease is usually centered in the cartilage, the surrounding bone and the membrane that lines the bone.

Arthritis and rheumatism are the most common, expensive and disabling of the chronic diseases. They occur twice as often as heart disease, ten times as often as diabetes and tuberculosis, and seven times as often as cancer. Medical care costs their victims over $100,-000,000 each year, not counting the additional sums spent on patent medicines, diets and mineral waters. About 97 per cent of all people past middle age develop some of the bone changes which are characteristic of arthritis, and one person in twenty suffers from some form of rheumatism. Fortunately, most of these cases are only annoying and painful, but do not cause deformity or serious crippling.

Do lightning rods really afford protection against lightning?

There are few natural occurrences more terrifying yet magnificent than the spectacle of a blinding flash of lightning accompanied by a resounding peal of thunder. This effect is due to the accumulation of large charges of electricity in the clouds which eventually break down the insulation of the surrounding air to flash across to ground or to another cloud. If the atmosphere were not such a good insulator of electricity, lightning would be greatly reduced in magnitude or absent altogether. Electricity would "leak off" before a great charge could build up. The use of lightning rods is an attempt to increase the leakage of electricity from cloud to earth, thereby preventing the build-up of a charge of any real magnitude. These metallic rods have sharp ends pointed skyward with the other ends planted firmly in deep, wet ground. Charges tend to congregate on the sharp pointed ends and, as a cloud passes over, a discharge of low intensity takes place. This discharge of electricity takes place so slowly and evenly that much of the cloud's electricity is neutralized. If lightning does eventually strike the house, its intensity is diminished considerably. Additional protection is afforded by the network of metal rods themselves which shield the house by carrying the large electric currents down to earth. Because of these factors, most authorities agree that lightning rods, when properly installed in appropriate situations, do give valuable protection against lightning.

How is wire made?
Wire is made by pulling a rod of the desired metal through successively smaller and smaller holes until the desired diameter is reached. A metal's ability to be drawn into wire is determined by its *ductility*. Such metals as copper, silver, gold and iron are said to be ductile. The device actually used to draw wire consists of a tungsten carbide die having a tapered hole of the desired size. The tip of the metal rod must be cut down to a size small enough to permit its insertion into the tapered hole. It is then fastened to a drum which pulls the wire through the die and then winds it up for storage. Wires of smaller size are made by drawing the wire through dies of the appropriate size.

How do rattlesnakes "see" in the dark?
Although rattlesnakes have pretty good eyes, nature has equipped them with organs, called "pits," which enable them to "see" in the dark. These are true organs, located on the sides of their heads, which respond to infrared (heat) rays. On the darkest of nights, the snake can "see" a mouse or squirrel by the heat of its body.

Another peculiar activity of snakes is their continual tongue-flicking. Each time the snake's forked tongue flicks out, it picks up odor-laden air which is then conveyed to smell organs inside the mouth. Snakes use this sense to follow the trail of prey after it has been injected with poisonous venom. The poison serves the dual purpose of killing the victim, and beginning the digestive process before the animal has been devoured.

Is the image in our eye upside down?
The image formed in our eye by the lens is actually upside down. Stated scientifically, the image is *real,* as opposed to the *virtual* image formed by a mirror, *inverted,* and *smaller* than the object viewed. If this is so, then why don't we see things upside down? The answer lies in our extremely complicated switchboard, the brain.

The eye itself really doesn't see anything; it merely projects an image on light-sensitive elements on the retina. These elements react to light rays by sending impulses to the brain where they are pieced together to form a picture of the objects viewed. This ability of the brain to reconstruct pictures from nerve impulses must be learned, and the learning process begins at birth. After years of practice, the infant learns to associate certain images with certain happenings. The upside-down character of images on the retina is just a part of this process. We learn the "ups" and "downs" of things by experience. If you have ever used a microscope, you are quite aware of this problem. At first you see everything "upside down," and following a moving microbe becomes a nerve-racking proposition. Lefts get confused with rights and ups with downs until you wonder if it is all worth while. You soon discover, however, that it's only necessary to move the slide the "wrong way" in order to follow the little rascal. Eventually the brain gets used to the new situation and automatically reverses your actions whenever you work with a microscope; the "wrong" movements seem natural and soon become habitual.

How long can a whale stay under water?
Whales are really mammals, and consequently have lungs instead of gills. How, then, do they manage to breathe under water? The answer is they don't. They must come up for air once every twenty minutes or so to replenish their supply of oxygen. The whale has nostrils at the top of its head, making it easy for him to get air when he comes to the surface. When he is submerged, special muscles close the nostrils and prevent the entry of water. The mouth is separate from the passageway which connects the nose with the lungs. This keeps water from reaching the whale's lungs even when his mouth is open.

Whalers depend on the animals' unique breathing methods to help locate them. The well-known phrase, "Thar she blows," calls attention to a fine spray rising from the surface like a geyser. This spray is really nothing more than air and water vapor which the whale has just exhaled. When the warm, moist air is exhaled into the cold atmosphere, the water vapor condenses, forming the columns of spray that people ascribe, incorrectly, to the spouting of water by whales.

How does our ear hear?

As you know, our ear consists of three parts: the outer ear, the middle ear and the inner ear. The outer ear collects and directs sound waves through the ear canal to the eardrum. This is a tightly stretched membrane, about three-thousandths of an inch thick, which separates the outer ear from the middle ear. Next come the three small bones of the middle ear called the hammer, the anvil and the stirrup. These bones transmit sound vibrations from the eardrum to the cochlea, a fluid-filled spiral passage in the inner ear. On the cochlea is the important organ of hearing called the *organ of Corti*. It contains a row of about 24,000 hairlike fibers which range in length from a 15th to a 170th of an inch. Scientists believe that the length of these hairs determines the frequency of pitch each will respond to. The shorter the hair, the higher the pitch. The shortest hairs respond to sounds of about 15,000 vibrations per second while the longer ones respond to sounds down to about 50 vibrations per second or lower, depending on the individual. When a given fiber vibrates in response to a sound wave, a nerve ending is stimulated and a corresponding impulse is sent to the brain. Our ability to distinguish, in this way, between tones of different pitch enables us to recognize voices and appreciate music.

What is meant by basal metabolism?

Have you ever noticed how hot a room gets, even in the middle of winter, when a large number of people are congregated in a small area? This is due to the heat given off by their bodies. A person walking slowly about a room gives off about 200 Calories, or 800 B.T.U. (British thermal units), per hour. This is quite a large amount of heat when we consider that an average five-room house in New York City can be heated in winter with only about 50,000 B.T.U. per hour, even on the coldest days. This means that about twelve or thirteen people, walking around the living room, can keep it warm even if the furnace is turned off. Where in the world does all this heat come from?

Think, for a moment, of all the vital processes that must go on incessantly even when one is asleep. The work of the liver and kidneys, the beat of the heart, the normal respiratory movements and many others must continue at all times. These functions demand a continuous supply of energy, and this energy is produced in our body by uniting the oxygen that we breathe with the food that we eat. The end product of this oxidation process is heat. The physiological processes involved are called *metabolism*. *Basal metabolism* is the heat output of the body, in Calories per minute, measured with the body at rest and without food for some time. It is measured, in practice, by a device called a respiratory calorimeter which measures the amount of oxygen we breathe in a unit of time and converts the answer to Calories per minute.

How was oxygen discovered?
Oxygen was discovered in 1774 by the English scientist, Joseph Priestley. Until that time, chemists believed that all combustibles contained a substance called phlogiston. When something burned, they said it was being dephlogisticated. Priestley, in his work on "different kinds of air," finally prepared some pure oxygen by heating red mercuric oxide until it decomposed. He found that substances burned much brighter and faster in this gas than in air. This led him to believe that it was present in air and was necessary for life. To prove this, he placed some mice in a closed jar until the oxygen was used up and the mice became unconscious. He then put some oxygen in the jar and the mice soon came to life and became active. Not to be outdone, he breathed some of the gas himself and noted that his breath "felt peculiarly light and easy for sometime afterward." He reported the results of his experiments to French chemist Antoine Lavoisier, who repeated them and named the new gas oxygen. The work of these two great scientists in the latter part of the eighteenth century marked the beginning of chemistry in the modern sense. Their work changed chemistry into an exact science and laid the groundwork for the great discoveries to come.

How are long tunnels ventilated?
Each of the two tubes of the Brooklyn-Battery Tunnel is thirty-one feet in diameter and almost two miles long. Automobiles and trucks passing through tunnels of this size would soon make them unusable if proper ventilation were not provided. In addition to giving off poisonous carbon monoxide gas, they would soon use up all of the oxygen in the tunnel and the gasoline in their engines would not burn. Three ventilation buildings were provided to insure adequate ventilation of the Brooklyn-Battery Tunnel. One is located at either

end and the third is located on Governor's Island, about midway between the two. Twenty-seven fans, some as large as eight feet in diameter, are used to force fresh air into the tunnel at sixty miles per hour. Ducts carry the air throughout the length of the tunnel and release it through ports near the roadway. Exhaust ports located at the top of the tunnel carry used air and poisonous gases back to the ventilation buildings where it is disposed of. It has been estimated that each vehicle making the 3½-minute trip gets four *tons* of fresh air for its passage through the tunnel.

How was petroleum formed?

Many millions of years ago, the sea existed over areas that are now dry land. As countless generations of sea animals decayed on the ocean floor, tiny globules of fat were formed and trapped by layers of sediment. Gradually, this sediment turned into shale, limestone or sandstone. Droplets of the fat drained through the rock, floating along until trapped by less porous rocks through which they could not pass. Natural gas then collected above the oil as the fatty matter decomposed further. The result of all this is petroleum; animal fat that formed millions of years ago and was trapped below nonporous rock formations.

The fact that oil originated in the sea explains why offshore reserves are so extensive. Some oil formations located far inland may have been shifted there by the powerful pressures within the earth which lifted or folded the oil-bearing rocks to their new location. Sometimes oil finds its way to the surface through faults in the rocks. This was the only source of the material for ancient peoples, such as the Hittites, who used it to grease the wheels of their chariots four thousand years ago.

How do fish breathe under water?

If we allow a glass of water to stand for a while at room temperature, bubbles appear inside the glass. These bubbles are a mixture of nitrogen and oxygen from the air. Although oxygen is only slightly soluble in water, enough is present to sustain the lives of fish. Instead of breathing through nostrils or lungs, fish are equipped to breathe through gills. The gill covers are bony flaps located just back of the head, one on each side. In the process of breathing, a fish closes its gill covers, opens its mouth and allows water to enter. When the mouth closes, the gill covers open and water is forced out through the gill slits. As water passes through, oxygen is absorbed by the many tiny blood vessels in the gills. These same blood vessels give off carbon dioxide which was produced in the life processes of the fish.

As an interesting aside, we might point out that fish also have nostrils. They are quite small but can be seen if you look closely on either side of the snout. They lead to a small sac where the olfactory sense is centered. In fish, the nostrils are used strictly for smelling and have no connection at all with breathing.

Can an iceboat sail faster than the wind?

The principle behind the ability of a boat to tack is this: The force always acts at right angles to the sail. This is true no matter what direction the wind is coming from. It's this fact that enables a sailboat to move at an angle into the wind. By properly orienting the sail with respect to the boat, and both with respect to the wind, it is possible to sail partially into the wind. Now let's see what limits the maximum speed a sailboat can make. Because of the principle of tacking, the boat is pushed by the wind in a direction *partially into the wind*. It follows that the faster the wind blows, the faster the boat will sail into that wind. The only factors impeding the speed of the boat are these:

1. The friction between boat and water (or ice, if an iceboat).

2. The friction between boat and wind (since it's going *into* the wind.

If there were no friction, *there would be nothing limiting the speed of the boat!* Friction is the only factor limiting the boat's speed! By proper design, the friction between boat and wind can be kept very low. The friction between sailboat and water is another matter, however. But in the case of iceboats, the friction between runners and ice is, by nature, extremely low. The ultimate speed of an iceboat, therefore, turns out to be extremely high—higher than the speed of the wind that pushes it! By keeping friction low, it's possible for an iceboat to sail faster than the wind, even though it is moving partially into the wind.

It's interesting to note that an iceboat sailing *with the wind* can't achieve such a high speed. As soon as the boat moves as fast as the wind, the wind ceases to exist as far as the sail is concerned. The wind can't catch up to the sail. Under these conditions, the wind can exert no force at all on the sail.

How can electricity clean the exhaust gases of a smokestack?

Have you ever noticed particles of dust dancing about in a sunbeam as it pours through a window? They refuse to settle down to earth even though the air in the room is very still. One theory to explain this action states that the particles all have an electrical charge

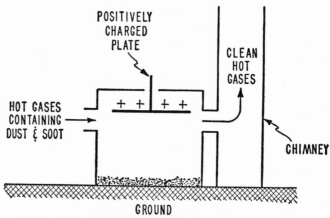

FIG. 14. CLEANING GASES ELECTRICALLY

Negatively charged particles of dust are attracted to the positive plate. Upon striking it they acquire a positive charge. They are then repelled to the bottom of the grounded chamber where their charge is removed.

of the same kind. This causes each particle to repel every other one, thereby preventing any from floating to the floor. If these charges can somehow be neutralized, the particles will no longer repel one another and ought to fall easily to the floor. A variation of this principle is used in dust precipitators which remove valuable chemicals from the exhaust gases of some industrial plants. The accompanying sketch illustrates such a device. The hot gases, dust and soot pass through a metal chamber before entering the smokestack. A metal plate is mounted from an insulator at the top of the chamber. This plate is connected to a large positive charge of

118

electricity. As the negatively charged particles of dust and soot pass beneath the charged plate, they are attracted to it because of the difference in polarity. Once they touch the plate, they lose their negative charge and acquire a positive charge. They are then repelled from the plate (since their charge is now the same as the plate) and accumulate at the bottom of the grounded chamber, where their charge is neutralized. The hot gases, cleaned of dust and soot, continue up the chimney.

Why are railroad tracks laid in relatively short lengths?

Steel rails are made in lengths to fit the gondola cars that are used to transport them. When they are laid, a space is usually left between rails to allow for expansion when the temperature rises. However, in some recent rail-laying operations, the rails were welded together into one continuous piece. The elasticity of the steel used was sufficient to overcome the stress caused by temperature rises.

Where do scientists look for antibiotics?

For many years it has been known that diphtheria, typhoid and other kinds of bacteria are destroyed in the soil. Bacteriologists now know that this is accomplished by certain kinds of soil bacteria which inhibit or suppress the growth of disease bacteria. Many research activities have been and are still being carried on to find varieties of soil bacteria that have this effect. The search is just beginning. Already hundreds have been isolated but time has not permitted many of them to be put to a clinical test. Soon many of these untried antibiotics will be used to treat diseases which have not as yet been conquered. When we consider the contribution of antibiotics in the span of time since World War II, it is easy to imagine great things in store for them in the future.

Why do explosives explode?

There are two general types of explosives: those which burn rapidly to form gases and those which decompose rapidly as a result of their inherent instability. In both cases, the tremendous pressures developed are due to vast amounts of gas suddenly released by the reaction. Black gunpowder is a typical example of the burning type of explosive. It is made up of a mixture of potassium nitrate, carbon and sulfur. It was invented by the Chinese thousands of years ago for use in fireworks. When ignited by a fuse, black gunpowder burns almost instantly because it contains both oxygen, in an available form, and combustible carbon. Tremendous quantities of nitrogen and carbon dioxide gas are liberated in the process and the well-known explosion takes place.

Some explosives, such as nitroglycerin and TNT, explode because they are basically unstable compounds. The element nitrogen is usually found to be the cause of this instability. Nitrogen is actually a very inert gas. It unites with other elements only with reluctance, and eagerly changes back to its gaseous form at the least provocation.

If nitrogen is caused to unite with oxygen and combustible materials, we have a veritable powder keg. When an explosive contains such a combination, detonation releases the nitrogen gas, and the oxygen unites with the solid combustibles to produce even more gas. The resulting explosion is extremely destructive.

Do emotional disturbances affect digestion?
Emotional disturbances are known to have an unfavorable effect on the digestion. In order to digest our food properly, certain fluids must be secreted in adequate quantities, and the motion of the stomach and intestines, called *peristalsis,* must be present to move the food along properly. Anger, fear and pain are known to have an adverse effect on both of these digestive needs.

There are many conditions which will arouse the emotions and thereby upset or retard digestion. Worry makes many of us lose our appetite; a rebuke during mealtime may make a child completely lose his desire for food; a sudden fright may give us a feeling of turmoil in the pit of the stomach. It is important that, as far as possible, our lives be so organized that our emotions are not aroused before or during mealtime. Children, for example, should not be scolded or punished at mealtime. Unpleasant experiences or situations should be avoided at all cost. Mealtime should be a well-organized period during which all activities are directed toward the fullest possible enjoyment of the meal.

Why does water start freezing at the top?
Most liquids contract, or get more dense, when they get cold. If water obeyed this rule completely, it would start freezing at the bottom since the most dense, and therefore the coldest, particles

would always settle to the bottom. What happens with water is quite unconventional in this respect. It contracts as it is cooled until 4° C. (slightly above freezing) is reached and then begins to expand again. This means that water is most dense at 4° C. Water either slightly above or below this temperature will be less dense and will rise to the surface. Since water at the freezing point (0° C.) is less •dense than water slightly warmer, it comes to the surface and freezes at that point.

What range of sound intensities can the human ear tolerate?
Most of us have heard the term *decibel* used in connection with the intensity or loudness of sounds. The decibel is a unit invented by engineers to help handle the enormous range of intensities that the human ear encounters in normal use. The following list will indicate the rating of sounds having a wide range of intensity.

Source	Rating in decibels
Faintest audible sound	0
Rustling of leaves	8
Whisper	10–20
Average home	20–30
Automobile	40–50
Conversation	50–60
Heavy traffic	70–80
Riveting gun	90–100
Thunder	110

The extreme range of sound levels listed above actually differs in energy content by a factor of one hundred billion to one! The use of decibels makes it possible for relatively small numbers to describe this great range of intensities. I'm sure you will agree that the ear is a truly wonderful device.

How does a Geiger counter measure radiation?
One of the simplest and most efficient methods for detecting the rays emitted by radioactive substances is a kind of vacuum tube invented by Hans Geiger and later perfected by W. Müller. In its essential form, the tube consists of a thin-walled glass envelope similar to an ordinary radio tube. It contains two metal plates and is filled with a small amount of a gas, such as argon. As in the case of a

neon lamp, the gas can be made to glow if the plates are connected to a high enough source of electrical voltage. When this high voltage is reached, the gas breaks down and permits a relatively large flow of electrons to take place between the plates. This flow is accompanied by the usual glow of the gas within the tube. When used in a Geiger counter, the voltage is purposely kept a bit too low to cause the glow discharge of the gas to take place under normal conditions. When a ray from a radioactive substance enters the tube, however, it collides with gas molecules, imparting enough energy to cause the glow discharge to occur. The current surge that results can be put through an indicating meter, or can be made to produce the ticking sound normally associated with Geiger counters. As soon as the ray has dissipated itself, the glow ceases and the tube is ready to receive another ray.

What are plastics?
The word "plastic" means "capable of being molded." Plastics are materials that can be shaped or molded while soft, and which harden into a relatively rigid condition. Such natural plastics as amber, tar, pitch and resins have been known for years, but only recently have synthetic plastics been developed. Most of these modern plastics are made by a process called *polymerization*. This is the chemical name for a process in which small molecules join together, like dancers in a conga dance, to form long, complicated molecules. Sometimes molecules of the same type join together, as in addition polymerization, while at other times different types of molecules unite, as in condensation polymerization. In either event, the end product is a very large and complicated molecule.

Plastics are sometimes called the "alloys" of the chemical industry because it is possible to make plastics according to specifications to meet certain requirements. There are the *thermoplastics*, such as nylon, which harden on cooling and which may be heated and cooled repeatedly without chemical change. Another group of plastics are the *thermosetting* plastics, which harden permanently upon heating and which cannot be softened again. The *casein* plastics, made from milk, do not fall into either of these classes. They harden at room temperature by a chemical change followed by drying. Many articles, such as buttons, billiard balls and chessmen, are made of casein plastics.

The use of plastics in the future is sure to increase because of their beauty, durability and utility. Even today plastics have taken over in our homes in the form of fixtures, carpets, draperies, jewelry, clothing, utensils, toys and building materials. Plastics must definitely be placed high in the list of materials considered important to man's progress.

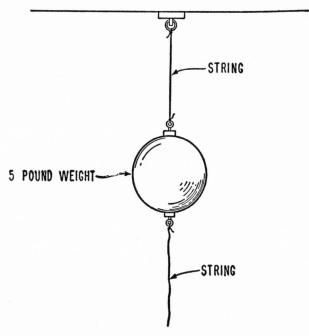

FIG. 15. INERTIA
A quick pull will break the lower string while a slow, steady
pull will break the upper string.

What is inertia?

If you are riding in an automobile and the brakes are applied suddenly, you have to brace yourself to avoid being thrown forward. Similarly, when an automobile is accelerated from a standing position, your body tends to press back against the seat. In both illustrations, your body has exhibited the property of *inertia*. This principle states that a mass at rest tends to remain at rest, or a mass in motion tends to remain in motion at the same velocity unless acted upon by a force. A person can be hurt seriously by jumping from a moving vehicle because he will continue to move at substantially the same speed until he hits the ground. Making a similar leap from a stationary vehicle presents no problem because his body has no forward motion. You can demonstrate that a body has inertia by a simple experiment. Tie two pieces of light string to a heavy object as shown in the accompanying sketch. Attach one string to a stable support. If you now pull down on the lower string, which string will break, the one above or the one below the weight? It all depends on how suddenly you pull. If you pull down suddenly, the lower string will break since the weight has inertia and takes too much time to move. If you pull down with a steady, slow force, the upper string will break since the force of the weight of the object and the force of the pull are both acting on the upper string at the same time.

There are many examples of inertia in everyday things. When you beat a rug, dust particles remain stationary while the rug moves swiftly away from them. Coal leaves a shovel because the coal tends to remain in motion after the shovel stops moving. Doors slam and baseballs fly through the air because of inertia. Everything that has weight exhibits this property and obeys its rules.

How much does the earth weigh?
Scientists have determined that the earth weighs about 7,000,000,-000,000,000,000,000 tons. Of course, a rather special weighing apparatus had to be devised to make this measurement. The principle involved is based upon the fact that any two objects exhibit a mutual attraction for one another. The force of gravity itself depends upon this attraction. In simple terms, the law states that two objects are attracted by a force which depends upon the mass of the objects and their separation. The larger the objects, the greater the force. The greater their separation, the smaller the force. In the scheme devised to weigh the earth, a small weight is suspended from a string and measurements are made to locate its exact position. A ton of lead is then brought near the dangling weight and the mutual attraction between the two causes the weight to be pulled ever so slightly out of plumb. The measurement of this displacement must be made with painstaking care as the small object is displaced less than one-

millionth of an inch! After this measurement is made, it is a simple mathematical problem to calculate the mass of the earth. Such calculations are based on the relative attraction of the earth and the one-ton weight on the suspended ball. The results of such calculations give the approximate weight of the earth in tons as a seven followed by twenty-one zeros.

Why are ulcers a man's disease?
Ulcers are most prevalent in the thin, wiry, nervous man engaged in some mental occupation. Of course, fat, phlegmatic men can and do have ulcers but such is not the usual case. One common trait in the ulcer victim is emotional instability. He takes life the hard way; he gets upset easily over disappointments, sorrows over injustices, and hides deep resentments under the brisk efficiency of a successful citizen. The ulcer sufferer strives continuously for excellence, and will not admit defeat even to himself. He is usually easy to get along with and makes a remarkably good adjustment in marriage.

Three out of four ulcer patients are men. In the case of women, they are almost invariably "career" women engaged in the same kinds of activities as men. If a female ulcer patient marries and settles down to life as a housewife, the ulcer usually disappears, often permanently.

Recent research seems to indicate that female sex hormones offer some protection against ulcers. On the basis of these findings, sex hormones have been used with some success in ulcer treatment. Work is continuing along these lines and may offer important findings in the future.

Can electricity flow in a vacuum?
We know that electricity can flow through wires, and that it flows through air in the form of lightning, but can it flow through a vacuum? The answer is yes, most definitely. Electricity flows through air with great difficulty because moving electrons, which constitute an electric current, constantly collide with particles of air. Each such collision impedes their flow just as in the case of a small boy running through a crowd. Air is evacuated from our ordinary radio tubes for just this reason. As air is removed from the space through which electrons must flow, they travel with increasingly greater ease.

When a very good vacuum is achieved, the number of collisions is reduced to an insignificant value and the flow takes place with greatest efficiency.

What is a scientific law?

The modern scientist has a skeptical attitude about the facts of his profession. He insists on firsthand observation and believes only the information that has been checked and double-checked in the laboratory. He has an "I'm from Missouri" philosophy which accepts as true only those concepts for which there is substantiating evidence. In the course of an experiment, a scientist may uncover certain new facts which do not fit into any existing pattern of events. After careful examination, these facts are considered with relation to all other known facts pertinent to the problem under investigation. As a result, scientists may formulate a theory or hypothesis which attempts to explain all of the related phenomena. Experiments are then devised to test or check this hypothesis. If it proves to be invalid, others are tried until one is found which seems satisfactorily to explain the events that are known to occur. When such a hypothesis stands the rigors of time and investigation, and when it can be used to predict facts other than those upon which it is based, it may be called a law or principle. Of course, there are really no such things in the strictest sense. A "law" is valid only until new investigations develop facts for which it cannot account.

How were sand and soil formed?

Geologists tell us that the earth was once nothing more than a big piece of solid rock. When it cooled slowly from the molten state, it formed one continuous sphere of the material we call igneous rock. Soon, however, the relentless forces of *weather* and *erosion* succeeded in breaking up the surface of this rock into the mantle of loose, fragmentary rock material that now covers the earth. The distinction between these two soil-forming forces depends upon whether or not the fragments are carried away by the forces that produced them. Erosion occurs when the material is carried away, while weathering takes place when the fragments remain in substantially the same place. Erosion commonly results from the action of moving water, winds and glaciers. Weathering is usually produced by the atmosphere with its changing temperature and humidity, although plants and animals break up rock to some extent.

The layer of loose material on the earth's surface is called *mantle rock* and includes everything from the finest particle of dust to the greatest boulder free to move. The solid material beneath the mantle rock is called *bedrock*. The mineral matter in soil and sand is nothing more than extremely small particles of mantle rock that were broken up by weather and erosion.

Can water be siphoned out of a leaking boat?

A siphon is nothing more than a self-operating tube or pipe which lifts a liquid and then deposits it at a point lower than the starting level. This qualification "lower than the starting level" is the key to the situation. As any waterboy knows, it's impossible to raise a liquid to a higher level without doing some work in the process. But a siphon can do no work since it is, by definition, a self-operating device. As a consequence, a siphon cannot empty a leaky boat since the water inside is presumably at a lower level than the water outside. But we might wonder why a siphon works at all.

Let's perform a simple experiment with two drinking glasses connected over the top by an inverted U-shaped tube. If we start with one glass and the tube filled with water, we find that water will flow through the tube into the second glass until the level is the same in each. If we then raise either one slightly, water will flow into the other until the levels are again equal. The important point is that water will always flow downward in the longer tube. As the longer column of water falls, it tends to leave a vacuum at the top of the tube. Atmospheric pressure then pushes water into the shorter side of the tube to fill this void. If we were to try to siphon water out of a leaky boat, our problem would be compounded since the direction of flow would be *into* the boat, a condition that would hardly be desirable.

How long does an explosion last?

Explosions are usually divided into two types depending on their duration. Black powder, for instance, is said to burn because the time for complete combustion of a charge can be measured in thousandths of a second. Explosives which operate on the *combustion* principle are used for propellants because they exert a uniform pressure which builds up slowly enough not to burst the gun barrel.

Dynamite explosions occur within millionths of a second. As you can see, dynamite decomposes a thousand times faster than black powder. Explosions of this kind, which take place in extremely short periods of time, are called *detonations*. If black powder is placed on a rock and ignited, it burns slowly and the gases escape harmlessly into the air. If the same experiment is tried with a stick of dynamite, a powerful explosion will result.

Why does air blow up a tire?

According to the *kinetic molecular theory*, all gases are composed of molecules which are in constant, random motion. Like bullets from a machine gun, they exert a constant pelting bombardment of all objects with which they come in contact. When we pump air into the inner tube of a tire, it inflates the tube, causing it to assume and hold its shape. This is due to the molecules of the gas, which continuously pelt the walls of the tube. Each molecule exerts a momentary force as it strikes the inside of the tube. The sum of all of these forces exerted by countless millions of gas molecules striking the walls of the tube constitutes the pressure exerted by the gas. Since there is a greater concentration of molecules inside the tube than there is outside, the net effect is to force the walls of the tube apart to the fully inflated position.

You may also have noticed that the air pressure in your tires tends to rise as you drive along hot concrete roads. This is caused by the heating of the molecules of air within the tires. Hot molecules have more energy than cool molecules and, therefore, move with greater speed. This causes them to hit the walls of the inner tube with greater force, with the resulting increase in pressure. You may be interested in knowing the actual pressure within your automobile tires. The pressure gauges normally used in service stations measure the difference between atmospheric pressure and the pressure within the tire. Since the atmosphere exerts about fifteen pounds per square inch and since you keep about thirty "pounds" of pressure in your tires, the total amounts to about forty-five pounds per square inch, or three times the pressure of the atmosphere.

What is a glacier?

Imagine a region so high that the mountaintops are covered with a thick mass of snow-clad ice the year round. As more and more ice and snow accumulate, they force their way down steep valleys toward lower levels. Following such a mass of snow-covered ice, we find that it slowly winds its way down, thinning out as it goes, until it suddenly ends. Careful observations indicate that such ice movements travel at a rate of several feet per day. Scientists call them *valley glaciers*.

Continental glaciers have their source in the extremely cold climate of the polar regions. For thousands of years, the only precipitation in such regions has been snow. It has accumulated and slowly changed to ice, forming a layer many hundreds and even thousands of feet thick. The only land visible through this sheet of ice is the tops of the highest mountains. This ice sheet moves outward from the center in all directions, pushed by the relentless pressure of its own weight. If it were not for this movement to warmer regions, both continental and valley glaciers would be much larger than they are.

Although we normally associate glaciers with ice ages of the past, they are very much in existence today. The Swiss Alps provide us with many majestic valley glaciers while the polar regions are covered with gigantic continental glaciers.

How much fuel does an atomic submarine use?

The U.S.S. *Nautilus,* the first atomic-powered submarine, was launched on January 21, 1954. It obtains energy for propulsion and other uses from a nuclear reactor which converts radioactive substances to lighter elements. The use of such a power plant enables a submarine to remain submerged for great lengths of time, since oxygen is not required by the power plant. From a knowledge of the probable efficiency of modern reactors, it is believed that the *Nautilus* could cruise around the world on the energy supplied by about two pounds of uranium. This is well below the critical amount required for an atomic bomb, so there is no danger of an atomic explosion. The second charge of nuclear fuel was placed aboard the *Nautilus* in April, 1957. The submarine had traveled sixty-thousand miles during a period of more than three years on a few pounds of fuel! This is equivalent to having used more than three million gallons of oil!

What causes the rainbow?

The rainbow we are all so familiar with is really on optical illusion resulting from the action of sunlight on small droplets of water. To see a rainbow, we must be between the sun and the droplets of water

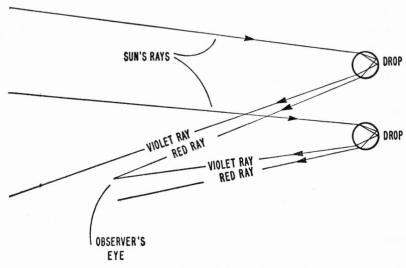

FIG. 16. PRINCIPLE OF THE RAINBOW

When light is reflected internally from the far side of a droplet, it comes out in different directions depending upon the color. A droplet at the top of the rainbow reflects only red light toward the observer, while a droplet at the bottom of the rainbow reflects only violet light in his direction.

with the sun at our back. If we are on level ground, the sun should be relatively low in the sky, not directly overhead. When these conditions are fulfilled, a rainbow spreads out before us in a beautiful semicircle of multicolored light. To understand where the colors come from, let's investigate the action of a prism on ordinary sunlight. You will recall that light striking a triangular prism is broken up into its component colors, from red at one end to violet at the other. It is this effect which causes the droplets of water to form the rainbow. Sunlight entering a droplet of water is broken up into its component colors just as though it had entered a glass prism. Some of this colored light is reflected internally from the far side of the droplet back in your direction. The important point here, however, is that the light comes back out of the droplet in different directions *depending on the color*. If you were to concentrate on one drop, say one at the top of the rainbow, you would see only one color, red. All of the other colors would be reflected up over your head and, hence, you could not see them. To see a violet color, you would have to fix your gaze on a droplet somewhat lower in the sky than the red one. In this case, all of the other colors would be reflected well below your feet. It is for this reason that the colors range from red at the top to violet at the bottom of the rainbow.

You can see only part of a circle because of the presence of the earth which obstructs the sun's rays. Airplane passengers, however, have been able to see the full circle spread out on a cloud with the shadow of the plane in the center.

Why does a rug feel warmer than a tile floor?
Have you ever walked across a tile floor in your bare feet and then felt relieved to step onto a rug that felt pleasant and warm? Although both materials were probably at the same temperature, one felt cold while the other was comfortable. You have also probably picked up an iron tool lying in the summer sun which was too hot to hold, yet its wooden handle felt only warm to the touch. In both instances, the senses reacted with complete indifference to the temperature of the objects involved. The reason behind these paradoxes lies in the facility with which each material conducts heat. Tile and iron are good conductors of heat while wool and wood conduct heat only with reluctance. When you step on a tile floor, the particles of tile in contact with your foot absorb heat and immediately pass this heat along to other particles and so on. Since tile is a good conductor of heat, a great deal of heat is removed from your foot in a short length of time resulting in the sensation of cold. A woolen rug, on the other hand, is unable to pass this heat along from particle to particle with nearly the rapidity of tile, hence it cannot absorb much heat.

Similarly, a hot iron tool is capable of passing a large amount of heat to your hand causing it to feel hot, while a wooden handle cannot. The outer layer of wood particles cools rapidly to the temperature of your hand. Since the inner particles of wood cannot replace the lost heat rapidly, the handle feels comfortable to the touch and not hot.

Why do so many atolls make excellent harbors?

Atolls consist of one or more coral islands arranged to form a sheltered, ring-shaped lagoon. They are believed to be coral reefs which grew around the fringe of sunken volcanoes. The crater provides the harbor's depth while openings in the ring of islands provide access to the ocean. Atolls of the Pacific, such as Midway and Wake, were famous during World War II as naval bases. Their quiet harbors provided excellent protection against winds and waves in the middle of a large and tempestuous ocean. Eniwetok and Bikini atolls attracted considerable attention as the historic sites of the early atomic bomb tests.

What are the chances of succumbing to mental illness?

Psychiatrists tell us that a neurotic person is one suffering from a mental disturbance which handicaps but does not incapacitate him. There are eight million such people in the United States today. There are over a million psychotics, those suffering from the most serious of mental illnesses. If we include alcoholism, mental deficiency and behavior disorders, we must add another 8.5 million. This adds up to a staggering proportion of the American population.

On the basis of figures like those given above, statisticians tell us that one out of every twelve persons in the United States will be hospitalized for a mental disorder in the course of a lifetime. In addition, one out of every four will develop a physical ailment attributable to emotional causes. In addition to these appalling figures, we are told that the incidence of neurosis is certainly on the rise. Experts attribute this rise to the tension, fear, uncertainty, anxiety and insecurity that exist in our world today to an abnormal degree. The problem of coping with these enemies of mental health is, evidently, a serious one. It would seem to deserve careful consideration on the part of each of us.

What is a mutant or "sport"?

A flower garden may contain flowers of the same variety which differ widely in size, color and general appearance. These variations are due to heredity and do not usually breed true in the next generation. Occasionally, however, an abrupt departure from type takes place in which the offspring is noticeably different from the parent. When this change is capable of transmission to future generations, the individual is known as a *mutant* or *sport*. The theory of mutation was first announced in 1904 by a Dutch botanist named Hugo De Vries as a result of experiments performed with ordinary garden primroses.

Scientists believe that mutations are caused by certain variations in the egg or sperm cells of the parent stock. It is possible that the chromosomes may become changed in character or that the genes may interchange their positions. It is also known that x-ray radiation is capable of producing mutations in some creatures. In any event, a mutant can be quite useful in the world of agriculture. When the first hornless calf was born, this characteristic was capable of being transmitted directly to future offspring, thus originating hornless cattle. Many desirable flowers and plants have been brought into existence by this process. Scientists will continue to search for the secrets of mutation, but all of us have an interest in favorable sports when they occur.

Why does a trumpet have to be retuned after it "warms up"?
Like all wind instruments, a trumpet becomes sharp after it has been played a while and must be retuned if it is to harmonize with other instruments. This increase in pitch is caused primarily by the heating of the instrument. As the instrument warms up, there is an increase in the velocity of sound through the air in its passages. If you're wondering what the speed of sound has to do with pitch, perhaps a simple experiment will be of help.

If you take two similar pop bottles and blow across the open end of each, the tones produced will be about the same. If you heat one of them under the hot water faucet and cool the other one, they will produce different tones. The hot bottle will sound about a half-tone higher than the cool one. The pitch produced by blowing across the bottle depends upon the *time* required for a sound impulse to travel from the open end down to the closed end and back again. Since sound travels faster in hot air than in cold air, the time required for the round trip is shorter in the hot bottle. This means that the hot bottle will produce more impulses or vibrations per

133

second, which the ear interprets as a higher-pitched sound.

The trumpet acts in much the same way. Although it is open at both ends, the pitch produced still depends upon the time required for sound impulses to travel back and forth from one end to the other. This period of time can be controlled by changing either the speed of sound or the length of the coiled tube. The latter method is the more practicable and is commonly used to tune such instruments.

How is milk homogenized?

Milk is a mixture of water, protein, fat, sugar and minerals. The fat or cream content amounts to about 4 per cent of the total and, being the lightest, tends to rise to the top. Homogenized milk is milk that has been changed mechanically to keep the cream uniformly dispersed throughout. In this process, milk is forced through very small holes under pressure while it is being pasteurized. This breaks the globules of fat into minute particles and disperses them evenly throughout the milk. Because of the extremely small size of the particles, they take a very long time to rise to the top; much longer than the average life of a bottle of milk in your refrigerator.

Homogenization of milk, contrary to popular belief, does not increase its nutritional value. The misconception probably arose because much homogenized milk is also irradiated, or fortified with Vitamin D. Such milk is especially valuable for children's use during winter months, when they get little sunshine. The added food value comes from the Vitamin D, however, and not from homogenization.

How does fuel ignite in a Diesel engine without spark plugs?

If you have ever pumped air into an automobile tire, you have probably noticed that the pump gets quite hot. At first glance we might suspect that the heat results from friction between the plunger and pump barrel. But this can't be so because the narrow rubber tube connecting the pump and tire also gets warm. The answer actually lies in a more unexpected direction. Molecules of air, or any other gas, are in continual motion. Since this motion is random in nature, we find that they sometimes collide with one another. Each such collision results in the conversion of a small amount of kinetic energy (energy of motion) into heat. In the pumping process, we crowd more and more molecules into a smaller volume. This increases the number of collisions per second and, hence, results in a greater rate of conversion of kinetic energy into heat. This is why the temperature of a gas goes up with increased pressure. The Diesel engine makes use of this principle by compressing air and vaporized fuel many times in the cylinder. When the pressure increases sufficiently, a high enough temperature is reached to ignite the mixture.

How do chameleons change color?

Although true chameleons are not often seen in this country, another lizard, called the "American chameleon," also has the ability to change color frequently. It is this American lizard that is usually sold at amusement places. He accomplishes his color changes by means of a number of tiny branched cells located beneath the skin. These cells contain pigments of various colors, and when the chameleon contracts or expands the cells, the position of the pigments is changed. The skin's color approximates that of the pigment closest to the skin. While most of us are prone to overrate the chameleon's ability to change color, they do, to a certain extent, tend to harmonize with the foliage on which they rest. Light and temperature are also important influences in the color changes, as are excitement and fright. The sum of all of these influences results in the lizard's color changes and the strange patterns that come and go on his skin.

Is there an easy way to clean tarnished silverware?

Chemists tell us that the black tarnish which appears on silverware is silver sulfide, a compound of silver and sulfur. In order to remove the tarnish, we must either remove the silver sulfide as a whole, or induce the sulfur to leave by itself. The former can be accomplished by mechanical scouring with one of the many silver polishes. The latter method, called electrolysis, is the more desirable because no silver is removed.

To clean silver by electrolysis, dissolve about two tablespoonfuls of baking soda (sodium bicarbonate) and two teaspoonfuls of salt in two quarts of water. Bring this solution to a boil in an *enamel* pan and place some bright aluminum foil in the bottom. Do not use an aluminum pan since it would become discolored by the chemical reactions involved. The pieces of silverware are then placed in contact with the aluminum foil. Be sure they are completely covered by the solution. Boil the solution for three minutes; then rinse the silver in hot water. The silver will be bright and clean. It all comes about as a result of a reaction which transfers sulfur from the silverware to the aluminum. This process of cleaning silverware does not remove or injure the silver in any way. It is effective, harmless and easy to do. It is particularly valuable in cleaning silver-plated materials since plated silver is softer than sterling and is more readily removed by abrasive polishes.

How far from a wall must you be in order to hear your echo?
The human ear can distinguish two sounds as separate only if they reach the ear at least one-tenth of a second apart. If the time interval is less, they blend in the hearing mechanism to give the impression of a single sound. If we shout across a lake or canyon, some of the sound energy may bounce back from objects on the other side, producing an echo. This is because of the distance involved which permits the original sound to die out before the reflected sound reaches our ear. Since sound travels at about 1,120 feet per second under normal conditions, it will travel 112 feet in one-tenth of a second. A sound must travel at least that distance to produce an echo. To hear an echo from a wall or other reflecting object, we must be located at least 56 feet from it—just far enough away to produce the required 112 feet round-trip distance.

Where do hot springs come from?
Hot springs are similar in origin to geysers. They result when subsurface water is heated by molten rock, or magma, deep in the earth. The difference lies in the quantity of water involved. Where the quantity is small, steam is generated and geysers are the result. When larger quantities are involved, the water merely heats up, producing hot springs. Water heated in this manner flows along until it reaches an opening in the earth, and such spas as Hot Springs in Arkansas and Warm Springs in Georgia are the result.

Iceland, as cold as it is, has about fifteen hundred hot springs, which produce water at 190° F. The people of the island drill wells to obtain additional hot water and today it is piped to heat hospitals, schools, swimming pools and homes. In Reykjavik, the capital city, most of the public buildings and private homes are heated in this way. A truly fascinating use of this water is the heating of greenhouses. In spite of its nearness to the Arctic Circle, Iceland is now able to produce tropical fruits in some quantity through the use of this unorthodox source of heat.

How does insulin help the diabetic?

In a normal person, sugar is converted quickly into energy to supply the body's needs. In the diabetic, sugar tends to accumulate because his pancreas can't produce enough insulin to change sugars and starch into energy. If he eats more than a very limited amount of such foods, they build up in his body, making him feel continually thirsty. In addition, the diabetic is always hungry since his body tissues are not properly nourished. The main symptoms of the disease are thirst, excessive urination and hunger.

In 1922, insulin was discovered by Sir Frederick Banting and Dr. Charles Best. It is now possible, by receiving injections of insulin, for diabetics to lead normal, productive lives. Before the availability of insulin, a sufferer of diabetes could exist only by starving himself. Even with the most severe of diets, the time came when treatment failed and the patient lapsed into a coma and died. How different things are today, when a diabetic often lives longer with the disease than, from statistics, he could hope to live without it.

Why is it easier to remember faces than names?

Psychologists tell us that there are three different ways to remember. To "remember" may mean to *recall,* as when we say, "I can't remember the answer." It may mean to *recognize,* as when we find that we are familiar with a television program being done for the second time. And it may mean to *relearn,* as when we try to "pick up" a foreign language studied in high school and long since forgotten. Our ability to recognize is measured by asking us to pick out material we have studied when it is mixed with other similar material. Our ability to recall is tested by asking us to reproduce learned material from the depths of our memory. Our ability to relearn is measured by determining the length of time we take to relearn as compared to the time it took the first time.

Of the three types of remembering, the act of recalling is the most difficult and makes the greatest demands upon us. Even though we may not be able to recall the answer to a question, it may still be possible for us to recognize it out of a number of alternatives. We say, "It is easier to remember faces than names," not realizing that we need only *recognize* the face, while we must *recall* the name from millions of bits of information hidden in our memory. If we were to reverse the situation—that is, ask someone to pick the name out of a group containing it, and describe the features of the face, or reproduce it—that person would say that remembering names is easier than remembering faces.

Why doesn't an explosive shell explode when fired?

Explosive shells are designed either for fragmentation or demolition purposes. Fragmentation shells have a time fuse which causes detonation of the charge in flight. Demolition shells explode only on striking an object. Each of these is equipped with three types of explosives: detonators, boosters and high explosives. The detonator is placed in the forward end of the shell where it can be actuated by the striker pin at the proper time. The booster, as its name implies, increases the force of the explosion started by the detonator. By this time there is sufficient energy available to explode the main bursting charge which usually consists of TNT or amatol. The explosives used in shells of this type do not explode when the gun is fired because the propellant provides a gradual build-up of pressure rather than a sharp blow. Actually, explosive shells are quite hard to detonate. On occasion, the detonator and the booster fail to set off the main charge and the shell becomes a "dud." It fails to explode even though it may have plunged many feet into the ground.

How is liquid air made?

Air or any other known gas can be liquefied by compressing it and simultaneously lowering its temperature. The reduced temperature requirement is necessary because no gas can be liquefied if the temperature is too high, no matter how much we compress it. The temperature to which a gas must be lowered before it can be liquefied is called its *critical temperature*. This is $-140.7°$ C. for air. In order to liquefy at this temperature, air must be compressed to 37.2 times atmospheric pressure. In practice, air is first freed of dust, carbon dioxide, water and other impurities. It is then subjected to a pressure of 3,000 pounds per square inch (about 200 atmospheres), while the heat caused by this compression is removed. Some of this cooled and highly compressed air is then allowed to expand in close proximity to copper tubes in which the main body of gas is located. In expanding, the gas becomes extremely cold. This expansion absorbs great amounts of heat from the confined gas and further reduces its temperature. Finally, a temperature is reached at which the compressed air begins to liquefy. After liquefaction starts, the process becomes continuous.

Liquid air consists mainly of nitrogen, oxygen and argon. These elements are obtained commercially from liquid air by making use of the fact that they have different boiling points. Nitrogen boils at −196° C. while oxygen boils at −182.5° C. If the temperature is held at −196° C., the gas leaving the mixture will be mostly nitrogen. When all of the nitrogen is removed, the temperature can be raised to −182.5° C. to boil off the oxygen. Argon and other rare gases are also obtained by this process.

At the temperature of liquid air (−190° C. when open to the atmosphere) the characteristics of many common substances change so radically that we find it hard to recognize them. Rubber becomes hard enough to shatter when dropped to the floor. Mercury becomes hard enough to drive nails into a piece of wood. A coil made of soft lead acts like an excellent spring, and the antifreeze in your car changes quickly into a solid. Observations such as these remind us that nothing can be thought of as merely a liquid, solid or gas. By changing the environment to which a substance is subjected, we can change its properties radically, and do amazing things in the process.

Where are the oldest mountains in the world?

Scientists believe that the oldest mountains in the world are located under our oceans! These mountains, unlike those on land, are not exposed to erosion and are believed to be many hundreds of millions of years old. In the relative calm of their undersea environment, they are expected to last as long as the earth does.

The longest mountain range in the world is the underwater mid-Atlantic range extending from the latitude of Iceland to the Antarctic region. Its shape generally follows that of the continents closest to it. The average height of this range is 7,000 feet with most of the peaks about 2,000 feet below the surface of the ocean. The island of Pico in the Azores is actually the top of one of these mountains. It totals 27,000 feet high, although 20,000 feet of it is beneath the

Atlantic Ocean. Most of these mountains may be hidden, but they are indeed real mountains; just as real as their familiar counterparts on land. And yet their existence was unsuspected until about a century ago, when soundings for the first Atlantic cable indicated their presence. The tallest undersea mountain of them all is Kauna Kea in the Hawaiian Islands, which is 13,784 feet above sea level and another 16,000 feet or so below the sea. In case you're wondering, this is just 643 feet greater than the height of Mount Everest.

How did the mermaid legend originate?
Mermaids have been the glamour girls of the sea for thousands of years, at least. They have been depicted in art as far back as the days of Babylon, about 1800 B.C. One plausible explanation for the origin of the mermaid legend is based on the sea cow. Its head is shaped much like that of the seal and its body is plump and somewhat fishlike. The significant feature, however, is the face which suggests a large, ugly human. When early navigators saw the sea cow stick its head up out of the water, they were probably struck by the animal's part human, part fishlike appearance. Since they had no optical instruments with which to make a closer study of its features, they took the easy way out and reported these creatures as glamorous mermaids.

How are synthetic diamonds made?
Diamonds are usually found in the vicinity of extinct volcanoes. Scientists believe that they were formed of carbon which was trapped in molten lava and subsequently cooled. A process such as this would subject carbon to extremely great temperatures and pressures. This might account for the differences in atomic arrangement between diamonds and ordinary carbon. In 1894 Henri Moissan, a French chemist, tried to make diamonds by such a procedure but the crystals that he made were too small for proper identification. It was not until 1954 that scientists succeeded in producing the first synthetic diamonds. The largest of these measured about one-sixteenth of an inch in length. They were made in a special press by subjecting carbon to a temperature of 2800° C. and a pressure of 800,000 pounds per square inch. The stones were yellow in color, consisting of 85 per cent carbon and 15 per cent ash. Chemical and x-ray tests indicated conclusively that they were real diamonds. Be·

cause of their imperfections, such diamonds will probably be used more in the manufacture of cutting tools than as gems. There would seem to be no theoretical reasons, however, to preclude the possibility of making perfect gems artificially in the future.

Do the particles of a solid move?
Although the molecules of a solid do not have the same degree of freedom as those of a liquid or gas, scientists believe that they do vibrate back and forth in relatively fixed orbits. Some solids, such as mothballs and camphor, evaporate when left open to the atmosphere. This would seem to indicate that the molecules are moving and that some of them manage to pop free into the air. Other evidence of molecular motion in solids has been obtained by experiment. If a piece of gold is clamped to a piece of lead, so that their flat surfaces are very close together, gold molecules move into the lead and lead molecules move into the gold. If pieces of zinc and copper are clamped together and left that way long enough, they may become joined together by a bond of brass. Examples such as these have been used to justify the theory that molecules of a solid do, in fact, move.

Why do some lakes have salt water, while others have fresh water?
The Great Salt Lake did not get its salt from the ocean, but rather from slow accumulation over the ages. As a matter of fact, the ocean's $3\frac{1}{2}$ per cent salt content is a poor second to the Great Salt Lake's 20 per cent. All of this material found its way into the lake in the form of dissolved mineral matter in the fresh water entering it. All such water contains a small amount of dissolved material obtained from the ground. The nature of this material depends upon the kinds of rocks found in the area. In regions of high rainfall, lakes overflow and lose about as much water as they gain, over a relatively long period of time. This means that mineral matter leaves at about the same rate as it comes in. Such lakes remain fresh since their salt content does not accumulate. In dry regions, however, evaporation may keep a lake's level low enough to prevent any overflow from its basin. In such cases, dissolved minerals must accumulate indefinitely since they cannot leave by evaporation as the water did. The water of the Dead Sea in Palestine contains over 24 per cent of dissolved mineral matter while Lake

Van in Turkey contains 33 per cent of dissolved salts! People float in these "heavy" waters with no exertion whatsoever, since the density of such water is greater than that of the human body.

It is interesting to find that while some lakes change from fresh water to salt, others do just the opposite. Some lakes, like Lake Champlain, originate as bays which have been cut off from the ocean. If the area is supplied with adequate rainfall, they are soon filled up as rivers and ground water pour into them. This continuous supply of water causes them to overflow, gradually flushing out the original salt water. In the course of time, such lakes become fresh, as will the Great Salt Lake if the climate of Utah ever becomes humid enough. If this ever happens, the lake will once again revert to the large fresh-water lake it was thousands of years ago.

What causes allergies?

The word *allergy* stems from two Greek words meaning "changed reaction." An allergic person is one who reacts to contact with certain substances in a manner that is different from normal, non-allergic people. An egg is an innocent and nourishing food to most people. Yet in certain men, women or even babies, a small amount of egg hidden in food will cause a severe rash, swollen eyes or an attack of asthma. Although these people may be normal and healthy in other ways, they are said to be "allergic to eggs."

The fundamental cause of allergies was discovered in 1910 by Sir Henry Dale, a British physician who later received the Nobel Prize for his work. Dr. Dale found that the body chemical, histamine, is the substance that touches off an allergic attack. Some defect in the allergic person's tissues evidently causes the production of too much histamine when in contact with allergic substances. The excess histamine then touches off an explosion of sneezing, coughing or wheezing. Scientists have not yet found out why some people are hypersensitive to certain substances. Although they know what happens, they don't as yet know why. Continued research will probably lead to a fuller understanding of the condition in the future.

What is polarized light?

In the terminology of the physicist, light is said to exhibit the characteristics of a *transverse wave*. To understand what he means, let's consider a simple analogy. Imagine a length of rope attached at one end to a fixed object such as a fence post. Let us pull up most of the slack and then jerk the rope up and down sharply. This motion produces a kink in the rope which moves along from hand to post. In this way, energy of motion from the hand is transmitted along the rope and dissipated in jiggling the fence at the other end. Now let's reconstruct the actions that took place. Our hand moved swiftly up and down, taking the end of the rope with it. As the resulting kink moved along the rope, each successive rope particle moved up and down in the same manner. The motion of the individual rope particles was always up and down. The wave motion (the moving kink in the rope) was parallel to the ground. Wave motion of this kind in which the particles move at right angles to the direction in which the wave travels is called *transverse*. Current theories state that light behaves in just this way.

To get a closer analogy with light, imagine that we now snap our end of the rope very rapidly in all conceivable directions. Instead of an up-and-down, wave motion, the rope exhibits transverse motions in all possible direction. Ordinary light of this kind, which seems to vibrate in all transverse directions is said to be unpolarized. Now imagine that we pass this light through a filter which removes all light but that vibrating in an up-and-down direction. Although our eye cannot distinguish the difference, this light is now vertically polarized and exhibits some peculiar characteristics. If we place a second filter behind and parallel to the first and rotate it, we find that a point is reached where no light gets through. Another point is reached where practically all of the light gets through. When the second filter is located to pass vertically polarized light, most of the light gets through. When it is rotated ninety degrees, it will only pass horizontally polarized light and, since this kind of light was removed by the first filter, no light at all can get through. This situation is analogous to a prisoner waving his arm through the bars. He can only move his arm up and down. If a second set of bars at right angles to the first is added, he cannot wave his arm through them at all.

A proposal to reduce the hazards of night driving would make all headlights and windshields of a material called Polaroid. Windshields would be polarized to accept light vibrating in one direction while headlights would produce light virbating in the opposite direction. Thus polarized light from the headlights of an approaching vehicle would be reduced considerably by the windshield of the other one. Ordinary light from the road, however, would pass through the windshield with only a minor reduction in intensity.

What is the difference between music and noise?
When we pluck the string of a violin, we produce a musical tone because the violin string vibrates to and fro at a regular rate. The same kind of vibration takes place when we strike a piano key or blow through a musical instrument. This kind of sound is pleasant and of steady pitch. A sound is musical if it is produced by an object which vibrates in a regular fashion. When we drop a glass to the floor, or allow water to flow from a faucet, we cause irregular vibrations. Slamming a door, hammering, and scraping our feet along the pavement are other examples of sounds produced by random, irregular vibrations. Such sounds are noise. The difference, then, between music and noise may be defined grossly in terms of the regularity of the vibrations that caused them. Music results from vibrations which occur at a constant rate for a significant period of time. Noise is due to random vibrations in which the ear can find no order or significance.

What causes dust storms?
In the summer of 1934, the Great Plains of the Midwest suffered the worst dust storms in the nation's history. These storms were caused by extreme drought and the removal of natural grasses from the plains. Rainfall between the Rocky Mountains and the hundredth meridian averages between ten and twenty inches per year. This is sufficient to grow wheat in the rainier years, but in dry years wheat will shrivel and die. The natural grasses, however, are adapted to live quite nicely with this amount of rainfall. Even in the driest years, their roots hold soil firm against the eroding force of the prevailing westerlies. Following the First World War, there was a great demand for wheat and beef in this country and abroad. Cattle were allowed to overgraze on the grasslands until the soil was bare. Wheat farmers moved past the hundredth meridian, plowing as they went, to raise wheat in a climate of marginal rainfall. When the great drought of 1934 hit the region, the crops withered away for lack of water. Since there were no grass roots to hold the soil

together, the western winds found a vast area of loose soil exposed to their action. Great "clouds" of once productive topsoil were blown high into the sky and eastward across the country in an almost continual series of dust storms. The result was a tragedy to the farmers of the Midwest. Much of this land has now been reclaimed by restoring the natural grass vegetation. Where farms remain, farmers have been educated to use the best methods of cultivation in order to conserve available rainfall. Contour farming has become commonplace and tree belts have been planted to reduce the force of the wind. Whether these and other efforts will be effective in preventing future dust storms, only time will tell.

How big are whale "babies"?

Whales are the biggest mammals of them all. We know that they are mammals because their young are born alive and are nourished by their mothers' milk. We know that they are big because one of them may weigh ten times as much as a full-grown elephant. Consequently, we would expect the young to be reasonably large, and such is the case. Whale calves are undoubtedly the largest babies produced by any kind of animal. There is one recorded birth in which an eighty-foot blue whale gave birth to a four-ton calf! No other animal can approach that record and it's probable that none wants to.

What is the difference between D.C. and A.C. electricity?

Electricity can be defined as the flow of negatively charged electrons in a closed loop or path. Suppose we connect an electric light bulb to a battery as in the accompanying sketch. Why does the battery produce an electric current and cause the lamp to light? What forces the electrons through the lamp? The answer to these questions lies in the fact that like charges repel one another. The negative terminal of a battery is so named because it contains an excess or reservoir of negatively charged electrons. The positive terminal, on the other hand, has a deficiency of these electrons. Perhaps a better name for this terminal would be "less negative" for there are really no positive charges involved. When the lamp is connected between the two battery terminals, the crowded electrons on the negative terminal force some of their number through the lamp and into the positive terminal where conditions are less crowded. This flow of electrons is due to the force of repulsion between them and is always in the same direction, from negative to positive. A flow of electrons that is always in the same direction through the external loop or circuit is called a *direct current*.

As you would infer from the above, an *alternating current* is one

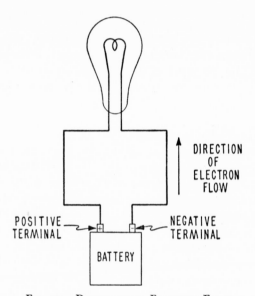

FIG. 17. DIRECTION OF ELECTRON FLOW
Electrons flow from the negative terminal of the battery, through the lamp, to the positive terminal. The negative terminal may be thought of as a reservoir of excess electrons. These force some of their number over to the positive terminal which has a deficiency of electrons.

in which the electrons change their direction in unison, alternately traveling in one direction and then the other. We can produce an alternating current by continually reversing the leads between our lamp and the battery. Since electrons always flow from negative to positive, the direction of flow *through the lamp* will depend on which way the lamp happens to be connected at any given instant.

If we reverse the leads several times a second, we notice that the

146

lamp dims noticeably during the lead-changing period. Since this flicker is objectionable, commercial A.C. systems accomplish this changeover sixty times a second, fast enough to make the flicker imperceptible. Of course A.C. electricity is not produced commercially by reversing the leads to a battery, but the results are about the same. Alternating current generators produce a smooth change in the direction of flow without the necessity of physically changing the connections of any wires.

You may wonder why power companies bother with A.C. electricity at all when D.C. seems so much simpler. Perhaps it will be adequate for our purposes to say that A.C. is easier to handle on long-distance transmissions and proves to be more economical in practice.

How were coral islands formed?

Man considers himself a great builder, yet his mightiest efforts are reduced to insignificance by the coral polyp, a relative of our everyday jellyfish. The Great Barrier Reef off the coast of Australia was built by corals and extends in length for 1,260 miles. It is 8,000 feet deep and varies in width from 7 to 100 miles. Quite a feat for a primitive animal that had its heyday about 350,000 years ago!

The corals secrete lime which eventually turns into *coralline*, a form of limestone. They live in large colonies with their bodies attached to one another and reproduce by growing buds which mature into new corals. Their skeletons remain in place, forming part of the reef. Scientists estimate that the rate of reef-building by corals amounts to about one inch a year.

Perhaps most remarkable of all is the fact that much of Mid-western North America, from the Gulf of Mexico to Hudson Bay, rests on coral formations built many millions of years ago. This confirms the belief that a large part of North America was once under water. We also know that reef-building corals can flourish only in a warm location, which substantiates other evidence that the Arctic regions once had a tropical climate.

What is the difference between heat and temperature?

All of us have read a thermometer at one time or another and know that temperature is the degree of hotness or coldness of something. Heat, on the other hand, is a form of energy—the quantity which enables machines to perform work. An abundance of heat means that a body (or the air, or anything else) is hot. The more heat it contains, the higher its temperature.

Back in the old days, people thought that heat was an invisible and weightless fluid, called caloric, which flowed from a hot body to a cold body like water over a dam. Although this theory has been disproved, our unit of heat, the calorie, is derived from "caloric" and represents the amount of heat needed to raise one gram of water one degree centigrade. This is the meaning that engineers and physicists use. When used in connection with nutrition and the energy content of food, it should be spelled in capitalized form (Calorie), in which case its value is one thousand times as great. The dual use of the term is confusing but there seems to be little that can be done about it.

Is gout caused by overeating?

Gout is a disease of middle age—one of the best known of the rheumatic ailments. Nineteen patients out of twenty are men, and in most of the cases there is a history of gout in the family. Contrary to popular belief, there is no truth in the notion that gout occurs only in wealthy, overweight people who indulge in rich food and drink.

A typical attack of gout starts with a sudden pain in the big toe, followed by swelling, redness and tenderness. It may also occur in the instep, the ankle, heel, hand, wrist or in several joints all at once. The acute stage usually subsides in from four to ten days but if not properly treated, it may lead to chronic joint disease. Patients can be kept comfortable with a drug called colchicine which usually relieves most acute attacks. ACTH and Cortisone also help severe cases, but do not cure the disease.

The cause of gout has recently been determined through the use of radioactive tracer elements by Dr. De Witt Stetten, Jr. of New York. He found that it results from an inherited defect which causes certain people to manufacture more uric acid than their bodies can utilize. This discovery will undoubtedly lead to an entirely new kind of treatment that will completely control this form of arthritis.

148

What is "hard water"?

Most of the water we use contains dissolved minerals which react with soap to form insoluble, curdlike substances. When the content of such minerals is excessive, the water is said to be hard. Hard water may be of two kinds, *temporary* or *permanent*. Temporary hard water contains calcium bicarbonate which decomposes when heated to produce calcium carbonate, ordinary boiler scale. It is this substance which collects within teakettles and hot-water pipes. Its presence in boilers is extremely undesirable because even a small amount obstructs the flow of heat and wastes fuel. Eventually, it may build up to a point where the efficiency and capacity of hot-water systems are seriously impaired.

Permanent hard water contains either calcium sulfate or magnesium sulfate (ordinary Epsom salt). They cannot be eliminated by heat, so unless removed by chemical means, they combine with soap to produce the curdlike substances mentioned earlier. Because of the inconvenience and high cost of using hard water, considerable research has been expended in finding inexpensive ways to soften it. Permanent hard water can be softened by the addition of washing soda followed by filtering out the insoluble substances which result. Temporary hard water is softened by adding lime to water and following the same procedure. In practice, both of these water softeners may be used at the same time.

A commercial water softener of a different kind is sold on the market under the trade name of "Calgon." Its chemical name is hexametaphosphate. When added to water before the use of soap, it reacts with water-hardening substances to produce new soluble compounds. These compounds remain in solution but do not react with soap to destroy its efficiency. It is not necessary, therefore, to filter them out before using the softened water.

Why do deltas form at the mouths of rivers?

You might at first suspect that the slowing down of rivers near their mouths is the cause of deltas, but actually, this seems to be only a small part of the story. The suspended material contained in river water consists of minute particles of clay, each of which possesses a negative charge of electricity. Since like charges repel one another, these particles of clay find it virtually impossible to settle to the bottom of the river. The lower particles repel those above and force them to remain suspended in the liquid. Such a mixture of small, charged particles and water is called a *colloidal dispersion*. The colloidal particles of clay and silt are not deposited until they come in contact with salt water which contains the positive charges necessary for neutralization. These positive charges are supplied by sodium chloride, the salt in salt water. When salt dissolves in water, it breaks up into two particles, a positively charged sodium particle and a negatively charged chloride particle. These particles, or *ions* as they are called, have independent freedom of movement within the water. As the negatively charged colloidal clay enters the salt water region, it obtains a positive charge from a sodium ion and, its charge now neutralized, it falls to the bottom of the river. During thousands of years, enough clay and silt settle in the area to produce a delta. The term "delta" arose from the similarity between the characteristically triangular shape of these deposits and the Greek letter delta (Δ).

What are "bone banks"?
Since 1946, when Dr. P. D. Wilson conceived the idea of preserving bones in a deep-freeze unit, bone banks have sprung up in hospitals all over the country. This system enables healthy bones from one person to bring health to another. The bone supply is obtained principally from amputated limbs, and from bones of healthy persons immediately after death. The cleaned bones are placed in a jar which is sealed and placed in the deep freeze. When a bone is required in surgery, it is removed from the deep freeze, thawed out in a few minutes and is then ready for use. After grafting, the frozen bone does not grow. Its value lies in stimulating the growth of new bone from the healthy bone on which it is grafted.

Why don't wells freeze in winter?
Experiments have shown that, at a depth of fifty feet, rock remains at the same temperature throughout the year. And more important, it assumes a temperature that is the yearly average for that particular location. In most parts of the United States, this average temperature is between 40° and 60° F. Well water is at the same temperature as the rock it permeates. This explains why it is refreshingly cool in summer and yet never freezes in winter. Of course, warm locations

like Florida and California have warmer wells than cold places in Maine and North Dakota. But none of these freeze because the average yearly temperature is above freezing. In polar regions, where the average temperature is below freezing, there can be no water wells since the ground water never thaws out.

What is ozone?
The oxygen we breathe exists in the form of molecules, each of which contains two atoms of the element oxygen. It is an odorless, colorless and tasteless gas. Whenever an electrical discharge takes place in air, however, some of the oxygen molecules are induced to change into a more active form of oxygen known as *ozone*. The pungent odor often present after electrical storms is due to the presence of ozone produced by lightning. It can also be noticed near x-ray and other electrical machines when they are in operation. Ozone molecules contain three atoms of oxygen instead of two, the third one having been annexed during the electrical disturbance. It is a much more vigorous oxidizing agent than oxygen, quickly destroying the elasticity of rubber and the pigments of dyes. It destroys germs very rapidly and tarnishes metals like mercury and silver with ease. Because of its germicidal properties, it is used in many European countries in the purification of water. It is quite unstable and, left to its own devices, will eventually change back into normal oxygen. Credit for its discovery must be given to M. van Marum, who, in 1795, first detected its presence in the vicinity of electrical machinery. The substance was given the name ozone by Schönbein (1840), who was the first to recognize it as a new and distinct substance.

What causes tidal waves?
Since tidal waves have nothing at all to do with the tides, scientists have generally adopted the Japanese word *tsunami* to describe this phenomenon. It is actually an unusually high wave caused by volcanic eruption or some other undersea disturbance. One great tidal wave, or tsunami, was caused in 1946 by an earthquake on the ocean floor near the Aleutian Islands. It traveled the two thousand miles to Hawaii in less than five hours and caused unbelievable damage. Houses and bridges were uplifted and hurled hundreds of feet. Over one hundred people lost their lives and property dam-

age ran into millions of dollars. As a result of this disaster, instruments have been developed to detect and chart the progress of tidal waves in the Pacific Ocean. The early warning that can now be given has resulted in a significant reduction in property damage and the loss of life due to such waves.

How does an amoeba eat without a mouth?

The amoeba is one of the lowliest of all animals. If you were to examine a drop of ordinary pond water under a microscope, you would have an excellent opportunity to study the amoeba in its natural habitat. It's merely a shapeless, jelly-like mass of protoplasm, the basic material of life, formed into a single, living cell. It can change its shape at will and moves about by thrusting out a projection called a pseudopod (false foot), and then squeezing all of its body into that projection. The amoeba "eats" its food merely by wrapping its body around small living plants, bacteria and other protozoans. The food is absorbed by the amoeba's protoplasm and is circulated by a movement of its protoplasm after which the food becomes living tissue. It "breathes" by absorbing oxygen through its surface. As oxygen combines with food, waste products such as carbon dioxide and urea are formed. These products are excreted mainly through the amoeba's surface. Thus, without any of the normal organs, and consisting of only one cell, the lowly amoeba is able to carry on all of the necessary life processes.

What is the highest sustained temperature that man can produce?

Hydrogen molecules are made up of two atoms of the element hydrogen rather closely bonded together. Dr. Langmuir, the noted American scientist, demonstrated that ordinary hydrogen molecules can be split in half into atoms by passing hydrogen gas through an electric arc. This cleavage is only accomplished, however, by supplying a tremendous amount of energy to the hydrogen. When such high-energy atoms of hydrogen are allowed to burn in air, the resulting combustion gives up all of this energy in the form of heat. The intensity of the flame depends not only upon the normal amount of heat produced by burning ordinary hydrogen, but also upon the larger amount of energy "loaned" to the atoms by the electric arc. As a consequence, the *atomic hydrogen torch* produces the highest sustained temperature which man has been able to attain. This temperature is estimated at 4,000° C.

Why do people like to sing in the shower?

Does a member of your family like to linger in the shower singing arias or intoning the musical scale? You may be interested in knowing that there is a basis in physical science for this tendency. It's called *reverberation*. The hard bathroom walls, often made of tile, reflect sound waves back and forth many times with little absorption. Before one sound dies out, another is added to it and the result, while not always too intelligible, is certainly big in *volume*. The mixing of a sound with those immediately following is called reverberation. In music, especially slow music, a given tone is held for an appreciable length of time. This means that reverberation adds to the power of a given tone resulting in a desirable (to the bathroom tenor, at least) expansion in volume.

Where intelligibility of speech is important, as in a meeting hall or schoolroom, too much reverberation can be confusing. For this reason, the ceilings of such rooms are often covered with sound-absorbing materials with many small holes spaced an inch or so apart. This reduces the reverberation characteristics of the room and makes conversation more enjoyable and understandable. Acoustical engineers tell us that a sound should die out, or become inaudible, about one second after the source has ceased sounding. Under these conditions, the average room sounds very good; not too dead, and yet not confusing.

What is a molecule?

According to the molecular theory, a *molecule* is the smallest particle of a substance that exhibits all of the properties of that substance. If we were to apply this definition to a grain of sugar, we would divide the particle in two, then divide one of these in two, continuing the process until a subdivision would produce particles which were no longer sugar. When chemists succeed in breaking a sugar molecule into smaller particles, they obtain not smaller molecules of sugar, but particles of a different kind—the building blocks of molecules. These smaller particles are called atoms and are really particles of the elements that make up sugar: hydrogen, carbon and oxygen. When hydrogen burns in air, two of its atoms unite with one atom of oxygen to form a molecule of water. When coal, which is mostly carbon, burns, an atom of carbon unites with two atoms of oxygen forming a molecule of the gas, carbon dioxide. In the illustrations just given, two *different* kinds of atoms united to form molecules. It is possible, however, for atoms of the *same* kind to form molecules under the right conditions. Gases such as oxygen, nitrogen and hydrogen normally exist as molecules in which two atoms are united. In the case of oxygen, an interesting difference in properties is noted when three atoms are joined to make molecules. This form of oxygen is called *ozone*, the substance that gives electricity its "smell." The fresh, pungent odor of air near electrical machinery is due to this form of the element. We can see from this illustration that ozone and ordinary oxygen are different substances even though they are made up of the same element. They are different because their molecules are different.

Why does the quarterback spin the football as he throws?
We have all seen a top or a toy gyroscope that stands upright without falling over. The phenomenon is familiar but the explanation is quite another matter. In fact, many mathematical treatises have been written to explain the existence of gyroscopic forces. Perhaps it will be sufficient for our purposes to state that the axis of rotation of any spinning body tends to maintain its original direction. The rotation of the body causes it to resist any effort to change that direction.

In the case of a football or rifle bullet, a spinning motion is used to prevent the object from tumbling over in flight. This rotation keeps the nose pointed in the initial direction, thereby minimizing wobble and reducing the effects of wind resistance.

What happens to gas in a gas mask?
The principal ingredient in most gas masks is a substance known as *activated charcoal.* It is made by blowing hot steam through charcoal, resulting in a very porous form of the material having an extremely large surface area. Activated charcoal will remove many times its own weight of poisonous gases from the air that passes through it. This removal is accomplished by a process known as *adsorption,* in which the molecules of a gas adhere to the surface

of solids in relatively thin films. Since activation increases the surface area of charcoal to a great extent, there is room for large quantities of gas to be adsorbed. Another feature of activated charcoal which makes it desirable for use in gas masks is its selectivity with respect to the gases that it will adsorb. It adsorbs very little oxygen while adsorbing extremely large quantities of many poisonous gases. Because of this selectivity, a gas mask will pass large quantities of oxygen into the lungs while filtering out practically all of the undesirable gases.

What are petrified forests?

Petrified forests consist of trees which time and circumstance have changed into stone. About 160,000,000 years ago there was a forest in Arizona. As the trees died, some of them fell into a stream and floated off to a near-by shallow sea. Repeated volcanic eruptions in the area then covered the trees with volcanic ash which contained the mineral silica. This mineral dissolved in the water and gradually turned the trees into various beautifully colored varieties of quartz. Such semiprecious stones as agate, onyx, jasper and opal abound in such quantity in Arizona's Petrified Forest that it is called the "Rainbow Forest."

There are twelve layers of petrified forests in Yellowstone National Park, and these were also formed by volcanic action. Smothered by volcanic ash, the trees could not decay, and instead gradually changed into stone. The Park contains two thousand feet of successive layers of volcanic ash and petrified wood which have been exposed to view by the erosive action of the Lamar River.

What is vitamin D?

Vitamin D is called the sunshine vitamin because sunlight produces the vitamin D compounds when it comes in contact with *ergosterol,* one of the substances found in human skin. In addition, ultraviolet light can be used to increase the vitamin D content of milk and certain other foods. This is accomplished by a process called *irradiation* in which the food is subjected to rays of ultraviolet light. Vitamin D is a mixture of at least four different compounds, all of which are known chemically as heavy alcohols. Nature stores it in the livers of fish, which provide a ready source of the substance for direct use as a food supplement. It is used by the body to aid in the assimilation of calcium and phosphorus, and a deficiency results in the disease known as *rickets*.

Why are sunsets so rich in reds and yellows?
When the sun is low in the sky, as at sunrise or sunset, the horizon is usually covered with beautiful warm colors ranging from the yellows to many shades of red. But if on such an occasion we were to transport ourselves instantaneously to a point several thousand miles to the west, we would find that the colors had miraculously disappeared. And furthermore, we would be back in the middle of the afternoon. As you probably have guessed, it is not the time of day, but rather the distance that sunlight must travel through our atmosphere that produces these colors. Our atmosphere tends to scatter out violet, blue and green light to a greater degree than the reds and yellows. When the sun is low it must travel through much more of the atmosphere than at noon and this selective scattering leaves a preponderance of reds and yellows for our eyes to see. The degree of redness on any particular day depends upon the character of the scattering particles through which the light must pass. Minute water droplets are particularly effective in scattering out the cold colors, thereby accounting for the beautiful coloration of certain cloud formations when conditions are correct.

The principle of selective scattering also explains why it is desirable to use red or yellow to penetrate fog or haze. Red neon lamps are particularly effective as aircraft beacons in overcast weather because their light consists of rays that are less susceptible to scattering. On the other hand, it's not scientifically sound merely to cover a white light with yellow or red glass. This may filter out the cold colors, but it doesn't increase the amount of fog-penetrating rays emanating from the lamp. Left to its own devices, the fog would have eliminated the blues and greens anyway. Neon and similar lamps are effective because they generate greater *amounts* of red light to start with; not because of the filtering out of useless colors. The only possible advantage obtained through the use of colored glass over a white lamp is the reduction in scattering of blue and green light which might be reflected back toward the driver as increased glare.

Why does the humidity affect our bodily comfort?

Have you ever felt chilly in a room heated to 75° or 80° F.? And have you noticed how uncomfortably hot a summer day can be at 80° or 90°? You are undoubtedly aware that humidity is connected with these situations, but let's try to get a more definite picture. To a great extent, these paradoxes are due to the rate at which our bodies can evaporate water into the atmosphere. Even though the temperature may be high, we can feel reasonably comfortable if the humidity is low. The air around us, under these circumstances, is far from saturated with water and is eager to allow perspiration to evaporate. When water evaporates from our bodies, it absorbs heat and cools our skin, making us feel at least comfortable. If the humidity is high, evaporation takes place slowly and less heat is removed from our bodies. This makes us feel uncomfortably hot even though the temperature may not be excessive.

Why do greenhouses get warm?

If you were to make a shallow box with a black bottom and a lid of glass, you would have an extremely efficient greenhouse. If such a box is placed in sunlight, the temperature will readily reach 300° F. or better. Just why does this temperature rise take place? Paradoxically enough, the answer lies in the fact that while glass is transparent to light rays, it is relatively opaque to the rays of heat. Light from the sun passes through the glass top of our miniature greenhouse and is almost completely absorbed by the nonreflecting black bottom of the box. All dark objects exhibit this tendency to trap light rays and convert their energy into heat. It is for this reason that most of us prefer, for summer wear, light-colored clothing which reflects most of the light rays it receives.

Since little light energy is reflected out of the box, it is absorbed by the box and changed into the longer wave-length rays that we call heat. Such rays cannot escape readily through the glass and the temperature builds up.

It is worth noting that light and heat rays are really quite similar in nature. You have, no doubt, noticed that the longer strings on a piano produce the lower-pitched sounds. It is conceivable that a piano could be made with strings long enough so that we could not hear a note at all. The energy coming from the string would be of the same nature, however, whether or not our ears were capable of responding to it. Light and heat rays are entirely analogous to audible and subaudible sounds. Light rays travel through glass and produce a sensation in our eyes. Heat rays can do neither of these things and, hence, remain trapped and invisible within the box.

What is the difference between wood alcohol and denatured alcohol?
Wood alcohol, or methanol, is a compound consisting of carbon, hydrogen and oxygen. It has a disarmingly pleasant odor but it is extremely poisonous, even if taken in small quantities. Blindness and even death may result frm drinking small quantities of it. It is used as a fuel and as a solvent for shellac. It was originally obtained by the distillation of wood, but today most of it is synthesized from carbon monoxide and hydrogen.

Denatured alcohol is a mixture consisting mostly of ethanol (the principal ingredient of alcoholic beverages), and certain poisonous and nauseating substances which are added to discourage its use as a beverage. Like wood alcohol, ethanol consists of carbon, hydrogen and oxygen, but in different proportions. The Federal government imposes a tax of sixteen dollars per gallon on pure ethanol, but denatured alcohol can be sold tax-free for industrial purposes. Rubbing alcohol is a kind of denatured alcohol mixed to a formula suitable for external use only.

Why does sound travel unusually well on a calm, clear night?
The speed at which sound travels depends to a great extent upon the temperature of the air through which it moves. For moderate temperature changes, the speed of sound goes up about two feet per second for each degree centigrade of temperature rise. If air were at rest and at the same temperature, sound would travel at the same speed in all directions. This condition rarely exists, however. On a warm day, the air layers near the ground are warmer than those immediately above. Since sound travels faster in the warmer layers, the bottom of a sound wave will travel faster than the top. This results in sound waves being bent away from the ground as shown in the accompanying diagram. Thus, most of the sound energy is bent up toward the sky rather than toward a distant listener.

On a calm, clear night the situation is quite different. The air is still and usually warmer than the earth. The air layers close to the ground are cooler than those above. Since sound travels slower in the cool air next to the earth, sound waves are bent down toward the ground. This makes the sound appear to travel farther than usual.

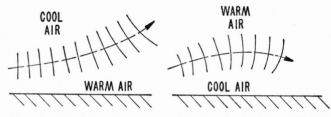

FIG. 18.   TEMPERATURE AND THE DIRECTION OF SOUND WAVES
Sound travels faster in warm air than in cool air. On a calm night, there is usually a layer of cool air near the earth and a layer of warmer air slightly higher. This causes the bottom of each sound wave to travel slower than its top. This results in a bending of the sound waves down toward the earth, and sounds seem to travel farther than usual.

What are fiords?

During past ice ages, the earth was much colder than it is today and a great deal of the earth's water supply rested in the form of ice on the polar regions. As a consequence, the level of water in the oceans was considerably lower than at present. In addition, many mountains in such places as Greenland and Norway produced great valley glaciers which cut deep gouges in the mountainside in their descent to the sea. After thousands of years, such glacial valleys had steep clifflike walls produced by the scouring action of their glaciers. When warmer times came, the polar ice began to melt and the ocean level began to rise. In addition, the glaciers became much shorter and no longer reached the lower levels of the mountain. As a result, many of the glacial valleys were flooded by the advancing sea. These partly flooded glacial valleys along seacoasts are called fiords. They usually extend well inland providing deep water for the navigation of large ships. The steep cliffs along their shorelines provide a natural condition for the formation of many waterfalls, giving fiords a reputation as the most spectacular form of coastal scenery.

Who was the first man to split the atom?

With the great current interest in nuclear physics, one might assume that it is a very new science. Quite to the contrary, the first assault on the nucleus was made back in 1919 when Rutherford succeeded in changing nitrogen into oxygen with only a bit of radium and a very simple apparatus. He constructed a metal chamber with a thin metal-foil "window" at one end. He then placed some radium near the end opposite the window and pumped the air out to form a vacuum. Rutherford knew that radium had the property of giving off alpha particles, which consist of two protons and two neutrons. None of these particles had sufficient speed, however, to move through the tube and penetrate the window. He proved this by placing a sheet of zinc sulfide outside the window. Any particles that succeeded in getting through the window would have caused light scintillations to appear on this zinc sulfide screen. Since none appeared, he knew that all of the alpha particles were dissipated within the tube. He then introduced a small amount of pure nitrogen gas within the tube. Light scintillations showed up immediately, indicating that something unusual was going on within the tube. It was subsequently proved that part of the alpha particle joined with nitrogen to produce oxygen, the next heavier element, and a proton that was left over in the process. This proton had a great amount of energy and therefore succeeded in getting through the metal window and reaching the screen. This was, indeed, a tremendous achievement. For the first time in history, man had succeeded in changing one element into another!

How did Portland cement get its name?

Although inferior cements were made in ancient times by the Romans and Egyptians, it was not until 1824 that a product was produced which resembled our modern cements in quality. It was invented by Joseph Aspdin, a bricklayer of Leeds, England. He called his cement "Portland" in order to benefit from the reputation enjoyed by "Portland stone," a building limestone quarried from the peninsula of Portland on the Dorsetshire coast. His cement was somewhat similar in appearance to Portland stone. Aspdin made his product by heating a mixture of limestone and clay in a kiln and pulverizing the resulting mass. Today, powdered limestone and clay are mixed in the proper proportions and heated until the mixture begins to melt. It then forms into hard masses, about the size of peas, which are called *clinker.* Grinding the clinker to a fine powder produces modern Portland cement. If a slower-setting cement is desired, 2 or 3 per cent of gypsum is ground in with the clinker.

Sometimes limestone deposits are found with the correct amount of clay already mixed in. Such rocks are called *natural cements.* Pennsylvania is noted for its extensive deposits of natural cement rock.

What does the sun look like from its farthest planet?

Pluto is a small planet located farthest from the sun. It is believed to be over three and a half billion miles from the sun, or about forty times as far away as the earth. If a person could live in the extreme cold that must exist on Pluto's surface, he would see the sun as a bright star without perceptible disk. To get an idea of the distances involved, let us imagine that we are able to communicate by two-way radio from the earth to either the sun or Pluto. If we were talking to someone on the sun, it would take eight minutes or so for our message to be heard. This is because of the time necessary for radio waves (or light waves for that matter) to travel the distance to the sun. A complete round-trip message would take twice that length of time, or about sixteen minutes. In communicating with Pluto, however, we would have to wait about ten hours for a reply. In spite of the tremendous speed at which radio waves travel, it would take about five hours for our message to reach Pluto, and another five hours for the answer to return.

How did the ancient Egyptians raise five-hundred-ton obelisks to an upright position?

Nowhere do we find more amazing feats of engineering than in the buildings and monuments that ancient peoples have left behind. Although these projects may seem simple in our day and age, how difficult their accomplishment must have been for these early peoples! Take the obelisks, for example. Imagine the Egyptian countryside dotted with these great masses of granite, often one hundred feet high and weighing up to five hundred tons each! How in the world did they erect these monuments? We can't be sure, of course, but the method probably depended on the use of an inclined plane. First, the base or pedestal was probably put in place and covered by a great flat-topped mound of earth. The sloping sides of the mound permitted the obelisk to be hauled, back end first, up the hill to the top. Then a hole was dug directly over the pedestal. The obelisk was then allowed to fall slowly into the hole. There is a good chance that a number of diggers may have been lost in this operation. When the obelisk was in a nearly vertical position, a rope was probably used to pull it upright. When this was done, there only remained the gigantic task of removing the great mound of earth.

Why is close intermarriage dangerous?

All of our physical characteristics are determined by genes which we inherit from our parents. Each cell in our body has two genes for every physical characteristic that you can think of. One has been inherited from the mother and the other from the father. Such factors as the color of hair, anemia, bleeding diseases, mental ability and countless others are determined in this way. Because of the mixing of genes throughout the ages, it is doubtful that any family possesses absolutely pure stock of favorable genes only. This means that even the best of families carry a few unfavorable genes from some black sheep or other in the family tree. Close intermarriage doubles these unfavorable genes so that they result in unfavorable characteristics in the descendants. This is because of the presence in both the sperm and egg cells of the same unfavorable genes. Although some intermarriages seem to have no undesirable effects, more often than not they run greater risks of passing on hereditary weaknesses.

What is the bottom of the ocean like?

Most of us take the ocean floor quite for granted since we expect to have very little to do with it anyway. But just what is it like down there? Scientists, with the aid of new and ingenious sounding equipment, have begun to learn quite a bit about the bottom of the ocean.

The average depth of the ocean is about ten thousand feet, although it varies considerably from point to point. It is extremely irregular, containing many mountains, valleys and gorges. Where the mountains are very high, their tops form our familiar islands.

As we would expect, the low points are continually being filled with all kinds of material. There are places piled high with countless millions of skeletons of various ocean creatures. One example is the *Foraminifera,* a creature so tiny that millions of them occupy less space than a cigarette. In spite of their microscopic size, they cover most of the ocean bottom to a depth of thousands of feet! In addition to this organic matter, rivers bring in great quantities of soil and sand which, in time, are changed into rock by the tremendous pressures. And in addition, some five hundred volcanoes add their clouds of ash to the air which eventually finds its way to the bottom of the sea. If we could stand the pressure down there, a trip to the bottom would surely be a fascinating experience.

Why are the freezing compartments of refrigerators placed at the top?

The freezing compartments of refrigerators are placed at the top in order to obtain the desired circulation of air within the enclosure. Cold air is heavier, or denser, than warm air. As air is cooled by the ice, it drops to the bottom of the refrigerator. This forces the less dense warm air to the top where it will come in contact with ice and be cooled. In this way, a constant circulation of air is provided which tends to maintain all of the enclosed air at nearly the same temperature. If the ice were placed at the bottom, on the other hand, very little circulation of air would take place. Air cooled by the ice would remain at the bottom of the refrigerator since, being more dense than warm air, it would have no tendency to rise. Warm air would remain at the top because it tends to rise. There would be a marked difference in temperature from top to bottom because of the lack of circulation of air. This temperature difference would be great enough to cause certain foods to spoil.

Why do eels breed in the Sargasso Sea?

Eels are amazing creatures. They appear at rivers as bright little things, only two or three inches long and no thicker than a match; they proceed up the rivers to the smallest brooks and ditches; they climb over rocks, damp grass and any other obstacles in their path; and they settle down in the most inaccessible inland waters imaginable. Several years later their eyes grow larger, their sexual glands expand, and their digestive glands atrophy. They leave their ınland homes, traveling over fields wet only with dew. Then they go downstream and disappear. What happens to them at this point had been a mystery for ages until cleared up by the tireless work of a Danish scientist, Johannes Schmidt. After years of work Schmidt discovered that all American and European eels disappear into the Atlantic and travel, probably at great depth, to a relatively small portion of the ocean known as the Sargasso Sea. It lies about midway between Puerto Rico and Bermuda and stretches eastward for some distance. Here they spawn at depths between three and fifteen thousand feet. It is probable that the parents die for they never return to land and are not again found in the ocean. The young eels come up to the surface and make the journey back to their parents' homeland (America or Europe as the case may be). The American eels make the trip in one year while the European species require three years for the longer journey.

Why do eels travel to the Sargasso Sea? The reasons are almost completely unknown. Conjectures on the subject are interesting however, and may even prove to be correct. Scientists believe that sexual hormones produced in the adult are responsible for a desire to move *with* the current (and hence to the ocean). The change of little eyes into big ones is probably responsible for their aversion to light and love of the ocean's depths. Their guidance to a certain portion of the ocean has been associated with temperature conditions in the Atlantic. During their period of migration, the eels show a marked need for a particular narrow range of temperatures —16° to 17° C.—their optimum temperature for reproduction. In addition, they require that this temperature exist at a depth of about fifteen thousand feet below the surface. The only place in the Atlantic that provides this environment is located in the Sargasso Sea—the Atlantic's great eel hatchery.

During the last ice age, it is probable that the conditions necessary for eel reproduction extended over a large portion of the Atlantic between the twentieth and eightieth parallels of west longitude. The American and European eels probably found it easy to locate spawning areas close to their respective homes. As the earth warmed up, however, the area of ideal environment became smaller and smaller until, today, only a small portion of the ocean seems acceptable. Since this area is much closer to America than Europe, the European eels have had to make an adjustment in their life cycle permitting them to make the longer voyage to their home continent.

Are there "man-eating" trees?

From time to time, travelers turn up a legend of the "man-eating" tree that is supposed to exist on the island of Madagascar. Investigation usually reveals, however, that the tree is to be found at some other part of the island from that at which the questioner is located. No authentic record of its existence has ever turned up. The nearest things to a man-eating tree appear to be the several plants which feed upon insects. The Venus's-flytrap has leaf blades forming two hinged surfaces. When an insect touches the sensitive hairs on either leaf, they snap together like the parts of a clamshell, imprisoning the insect. In the pitcher plant, the leaf is formed into a cuplike pouch. The inside of the pouch is covered with stiff, downward-pointing hairs. These make it easy for the insect to climb down into the pouch, but difficult for him to get back out. The victim eventually falls into the fluid located in the bottom of the pouch and is digested. As for man-eating trees, there probably aren't any.

Why don't spiders get stuck in their own webs?

As a youngster, the question of spiders and their sticky webs was always a mystery to me. Why in the world were they immune to the sticky threads that seemed to adhere so well to their enemies? The answer is quite simple. They aren't. They are intelligent enough, however, to provide for their own safe conduct through the web. It's accomplished in this manner. The spider spins the original web of dry, nonadhesive threads. When the web is just about finished, the spider goes over most of it again, this time with a sticky and gummy kind of thread. It is this thread that catches trespassing insects. The spider is careful, however, to leave certain areas free of this sticky material so that it may run up and down the web with no danger of becoming stuck.

Why isn't it necessary to pump water out of an artesian well?

An *artesian well* is an example of water seeking its own level. If we drill deep enough in some locations, we will finally come to a layer of impermeable rock through which water cannot pass. Beneath this layer there exists a stratum of porous rock material filled with water. Further down, a second layer of impermeable rock prevents this water from sinking any lower. The water-bearing material is sandwiched, so to speak, between two impermeable layers of rock. If this "sandwich" is tilted, so that it finally comes to an end at the earth's surface, water will enter the porous material and fill up almost to the top. The typical artesian well is located somewhat below the water level in the porous layer of rock. Since water seeks its own level, it will be forced up and out of the well with considerable force, due to the pressure exerted by the water in the tilted sandwich of rock.

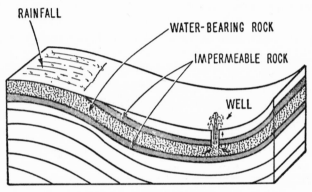

FIG. 19. AN ARTESIAN WELL

An artesian well is one from which water flows spontaneously. It is an example of water seeking its own level.

What are pearls made of?

When a grain of sand or some other impurity gets between the shells of a certain mussel or pearl oyster, the mussel covers the irritating intruder with successive layers of *calcium carbonate,* forming a pearl over a period of years. With today's modern methods, it's possible to place "pearl seeds" in the shells of certain mollusks and thereby take pearl production out of the hands of chance. Whether produced by artificial seeding or not, the perfect pearl has an iridescence that is unequaled by that of any other gem.

Calcium carbonate, which is the basic material of pearls, is also found commonly in other forms. When water seeps through the ground, calcium carbonate goes into solution forming a kind of "hard water." When this water finds its way into caverns and caves, it evaporates, leaving stalactites and stalagmites. If hard water of this kind is used in heating systems, the calcium carbonate is deposited in the pipes in a form known as "boiler scale." Ordinary limestone and marble are forms of calcium carbonate formed by pressure and time from the skeletons of minute sea animals.

The finest and most valuable pearls in the world are grown in the tropical waters of the Persian Gulf, chiefly around the Bahrein Islands. Practically no pearls of any value are ever found in the edible oysters of North America. Although pearls are found in these oysters, their limited luster leaves them practically worthless. In the event you do have a string of pearls, never leave them in an extremely dry location. Pearls deteriorate if left in dry safe-deposit boxes because they require moisture for the preservation of their luster.

Where does each day of the year begin?

Let's take an imaginary trip around the world, starting from New York City, and traveling westward across North America, the Pacific Ocean, Eurasia and finally the Atlantic Ocean. To make it even more interesting, let's fly in an airplane at about 750 miles per hour, just fast enough to keep up with the sun. If we start our trip at 12:00 noon on July 6 (we could have picked any other day), we will fly completely around the earth without ever seeing the sun go down. Quite to the contrary, our speed is just about right to keep the sun directly overhead for the entire journey. After circling the earth, we will arrive back at LaGuardia Airport at 12:00 noon the following day, July 7. But where did the new day come from? Where did it begin? New Yorkers spent the 24-hour period in normal fashion, witnessing afternoon, night and morning. Those of us in the plane saw none of this, since it was noon for the entire trip.

In order to obviate these and similar difficulties, the international date line has been established at the 180th meridian of longitude. When a traveler passes the date line in a westerly direction, the day automatically changes to "tomorrow." Assume that we reached the date line at 12:00 noon (since it was always noon on our journey) on July 6. As soon as we crossed over, it became 12:00

noon on July 7. It was also 12:00 noon, July 7 when we landed at New York since we kept up with the sun's pace through the heavens. Our journey, therefore, took exactly one day, thanks to the international date line.

When a traveler crosses the date line in an easterly direction, the opposite sort of thing takes place, and the day changes to "yesterday". An eastbound traveler, reaching the date line at 7:00 P.M. on December 25 finds himself moved back to 7:00 P.M. on December 24. Not only is he theoretically entitled to two sets of Christmas gifts, but he should also get two Christmas day dinners. Lady travelers are cautioned against crossing the date line in this direction on their birthdays since their age would then be increased by two instead of one for the entire transaction.

Why do snakes continually flick their tongues?

Tongue-flicking in snakes is usually taken as an ominous sign by most of us, who really don't care too much for the slithery things anyway. But the basis of the habit is really quite simple and inoffensive. Snakes flick their tongues to aid in the process of smelling. Although they have no noses, snakes do have an olfactory organ, known as Jacobson's organ, which is located in a cavity far forward on the roof of the mouth. Since air cannot be passed over the organ in the breathing process, the tongue does the job by bringing in samples of air located in front of the head. Tongue-flicking is merely a part of the snake's method of smelling the world around it.

How can we tell the age of a tree?

We can tell the age of a tree merely by counting the concentric rings of its cross-section. This would seem to be an adequate contribution to man's store of knowledge, but its importance is dwarfed by other marvels that scientists are beginning to find in these rings. The rings of each tree are really a diary in which its life history is inscribed in minute detail. They have given us a continuous history extending back in time over three thousand years! This is how it's done. Each ring indicates the annual increase in girth of the tree. The rapid growth of spring is light in color and texture, while summer's growth is darker and coarser in preparation for winter's dormancy. During periods of drought, the growth is slow and blurred. Sometimes calcium and other minerals accumulate in the wood. During certain periods of years the rings may crowd together, while at other times they spread apart to disclose a cycle of rapid growth. These and other variations in ring structure all have significance in interpreting the environment of past centuries.

While other factors must be considered, growth rings seem to be most closely related to available moisture. Years of great growth were years of high rainfall. Using this information, Professor Huntington traveled across the world studying tree rings. They helped him trace the climatic changes that have affected mass migrations of peoples and the rise and fall of empires. His investigations extended back in time to a giant Sequoia that began life in 1305 B.C.

Another valuable investigation was carried on by Professor Andrew E. Douglass of the University of Arizona. His studies of the logs used in the construction of the pueblos of the Southwest unearthed a wealth of detailed information. Although the illiterate tenants of these ancient buildings could leave no written records, the dry climate of the region has preserved many diaries of the past in the form of trees felled for lumber. Dr. Douglass correlated the rings found on these beams with tree-ring records of other areas. He found that one roof beam had been put in service in the year 1370. Another started life in 1075 and was cut down for timber in the year 1260. One of the buildings, Pueblo Bonito, was built in 919. The wood used in its construction carried the diary of rainfall for that region back another twelve hundred years.

By studying tree rings scientists have been able to determine three things: the relative rainfall in any given year; the cycle of sunspot activity as it occurred in the past; and the more general changes in climate that have taken place on the earth. Perhaps of equal importance is the insight that tree rings give them into the future. By knowing accurately what has happened to our environment in the past, scientists may be able to predict what is going to happen to it in the future.

How do insects find their way home?

It's highly improbable that insects can memorize their surroundings and use this information to guide them in their travels. With his limited brain power, it's inconceivable that a bee could remember every landmark between his hive and the flower fields. A more probable answer was provided in 1911 by Santschi, who called his discovery *light-compass orientation*. Santschi showed that insects guide themselves by their relative position with respect to the sun. Thus, insects were the first to make use of celestial navigation. You can prove this by watching the direction of motion of an ant that seems to be "going some place." A meandering ant won't do for this experiment since his motion will not make a constant angle to the sun. After noting his direction, place a small black box over the ant and leave it there for an hour. During this period of time, the sun will have moved fifteen degrees in the sky. Remove the box at the end of an hour and you will see the ant continue on his interrupted journey in a direction that is fifteen degrees removed from his original path. If this behavior seems complicated to you, just imagine the ant's problems on the return trip! Not only must he reverse his angles, but he must interpose right and left to make it back home. Despite the difficulties involved in this type of navigation, the ants, wasps and bees all seem to use it.

What is the difference between organic and inorganic substances?
For simplicity, chemical compounds have been divided into two classes: *organic* and *inorganic*. Organic compounds are those which contain the element carbon—the same carbon that we find in lead pencils, lampblack, coal and diamonds. Inorganic compounds include everything else. Prior to 1828, it was thought that only living things could produce organic compounds, and it was believed that their secret was somehow connected with life itself. But in 1828 a German chemist, Friedrich Wöhler, succeeded in producing the organic chemical, *urea,* synthetically. Since that time, chemists have synthesized many other compounds which previously had been obtained only from nature. The result, of course, has been an amazing increase in the number of products available for our use. There are at least 300,000 carbon compounds, or about ten times the number of all the inorganic compounds combined. The field of organic chemistry has become so fantastically complicated that scientists must specialize in a single branch of it, such as cellulose, petroleum, vitamins, fats, sugars, etc. They make no attempt to know all there is to know about branches other than their specialty. Typical examples of organic substances include foods, enzymes, vitamins, vegetable medicines, many acids, olive oil, linseed oil, flavoring materials, and many more. Fermentation of

organic substances gives us such products as alcohol, lactic acid and vinegar. Coal gives us coal tar, which is the starting point for the manufacture of at least a hundred organic products including plastics, disinfectants and explosives. Other products from coal include aspirin, ammonia, saccharin, food preservatives and many dyes. All of this is possible because of organic chemistry, the study of carbon and its compounds.

How deep do fish live?

The greatest depth at which fish have been caught is 6,000 meters or over 19,000 feet below the surface of the ocean! The deepest point in the ocean is located near the island of Mindanao in the Philippines. It measures 31,600 feet deep. It's quite possible that fish may be found at this depth one day.

Another striking fact about the ocean's depth is the complete absence of light. The last traces of light disappear at about 2,800 feet. How then, you might wonder, do our deep-sea friends manage to find their food? Nature, with her continual flair for the unusual, has solved the problem by providing many of these fish with spotlights—about 10 per cent of them in fact! These light-producing mechanisms vary in size and development from simple luminescence to complicated and powerful organs with focusing lenses, reflectors and on-off switches! Light organs are usually cup-shaped depressions below transparent layers of skin. The depressions are usually lined with opaque layers of black or red pigment covered with reflecting fibers; over these fibers are located the light-producing cells. A lens is located to collect this light and focus it out in a beam. Many of these fish are able to turn and move their light organs by means of special muscles, and most of them are able to turn the light on and off at will.

The light found in such fish is sometimes produced by luminous bacteria living between, or in, the fish's light cells. This close association of two entirely different creatures is known as *symbiosis*. Each partner profits from the arrangement: the fish get their spotlights, and the bacteria get an easy and protected existence. Some fish, however, are able to produce light without the aid of *symbiosis*. In either event, the chemical process involved is similar to that encountered in the firefly, and has been discussed in a section on that subject. If you're wondering how the less fortunate, lightless fish manage to catch their prey, perhaps they eat the fish with lamps.

Why do we see different stars in summer than in winter?

Some of the summer constellations are different from the ones we see in winter. This occurs because we can see stars only from the nighttime side of the earth—the side away from the sun. The earth rotates in its orbit about the sun and in summer and winter it is at opposite ends of this orbit. Since we see stars only from the darkened side of the earth, a portion of the summer sky is different from that of winter. This results in our seeing different groups of stars from one season to the next.

What is hibernation?

Without going too much into detail, the true hibernator changes in autumn from a warm-blooded animal into a cold-blooded one. The animal's body temperature drops to a point close to that of the environment. In some bats, for example, body temperatures have been measured as low as 32° F. below zero! During this period of lethargy, the hibernating animal experiences a slowing down of the heartbeat and respiration. High concentrations of carbon dioxide in the blood no longer induce faster breathing. Control of body functions is not absent completely, however, since extremely low temperatures result in increased heartbeat and heat production. The energy needed to maintain the animal during hibernation is supplied by fat stored during his active period.

Recent experiments have shown that hibernation is induced in most animals in northern latitudes by a combination of fasting and cold weather. By combining these two factors, fat dormice have been held in hibernation for as long as a year. But whatever its true physiological nature, hibernation provides practicing species with the ability to live in zones that would otherwise be just about uninhabitable.

Can a kitchen be cooled by opening the refrigerator door?

Have you ever walked past the refrigerator on a sultry afternoon and entertained the idea of releasing its treasure of cold air into the room? Appealing as this idea is, don't give in to it for the benefits will be short-lived. Scientifically speaking, there really is no such thing as "cold." Cold is simply the absence of heat. Something is cold because it contains less heat than a hot object. Ice is water that has had some of its heat removed. Our bodies sometimes feel cold because heat is being removed from them at a rate that is greater than normal. Similarly, the inside of a refrigerator is cold only because heat has been removed from it. A refrigerator is, after all, merely a mechanical device that pumps heat from one place to another. The inside of a refrigerator is cold because heat has been pumped out of it. Since the cooling mechanism has no ability to store this heat, it releases it into the kitchen. If the refrigerator door is left open, an endless cycle begins. Heat from the room is absorbed by the freezing unit of the refrigerator. It is then acted upon by the refrigeration equipment and pumped out into the room. It then goes back into the freezing unit, and so on. Oddly enough, opening the refrigerator door would soon cause the room to get warmer! Once the ice has melted, the heat of the refrigerator's electric motor would add to the heat in the room causing the temperature to rise! Most small, room-type air conditioners must be placed in a window to provide for the disposal of heat. Such devices throw the heat outside. Larger air conditioners use cold water from wells or other sources for the disposal of heat. A remarkable characteristic of some air conditioners called *heat pumps* is their ability to reverse the flow of heat. In the summertime they pump heat out of the house, leaving the inside cool and pleasant. In cold weather, they can be reversed to pump heat from the outside to the inside, thereby heating the house. If this sounds incredible, keep in mind that everything contains a certain amount of heat. Even snow and ice can be cooled to a lower temperature by taking heat out of them. Even if the temperature is ten degrees below zero outside, the reversible heat pump can take some heat from the outside air and pump it into the house. Perhaps, someday, many of our homes will be heated this way.

Can insects tell time?
Oddly enough, this is a difficult question to answer with certainty. Experiments with wasps seem to indicate that they pay little attention to time, but depend upon their environment to control their acivities. There appear to be two factors which determine their awakening: temperature and light. If either environmental factor

is below a certain limit, they will not wake up. If the lights are turned on, however, and their nests are warmed up, the wasps will wake up even in the middle of night. If either the temperature or the light level is reduced, they return to their nests and go back to sleep. This would seem to indicate that wasps pay little attention to the passage of time. Ants, on the other hand, only add to the confusion. They make use of nuptial flights to increase the number of colonies of their species. In such flights, thousands of males and young queen ants take flight after which the queens establish new colonies of their own. Investigations have shown that all colonies of a given species have their nuptial flights on the same day. This is true even if the colonies are separated by tens or hundreds of miles. And to make matters more disturbing, they seem to pick the same time of day for this activity. This seems a little too clever, even for ants. As for the time-telling ability of insects, a final verdict will have to await the results of future investigations by ever curious scientists.

How can seemingly healthy trees be hollow?

If you ever have an opportunity to examine the stump of a tree recently sawed down, take a close look at its cross-section. You will probably find that the central part of the trunk is denser and darker than the outer section. This central portion of the tree is called the *heartwood*. It is the strongest part of the tree and makes the best lumber, but . . . it's dead. Its once living cells have long since died, their tissue having been toughened and weatherproofed by the addition of resins. The heartwood of a tree acts as a central core of strength, helping to provide support for the rest of the tree. The living wood surrounding the dead core is usually more porous and lighter in color. It contains the maze of passageways needed for the flow of water and sap from the root system to the leaves. This portion of the tree is called the *sapwood* and it usually extends only a short distance into the trunk. In a peach tree, only the outer three or four yearly growth rings are really sapwood. Even so, the growth of a tree is restricted to the outer layer of the outermost tree ring exposed to view. This layer, called the *cambium*, is only one cell thick and it's here that the process of creating new tissue takes place. The cambium consists of microscopic cells that are constantly dividing. Cells that split off toward the center of the trunk become part of the sapwood, while those splitting off to the outside form the inner bark. The cambium is in a continual state of growth, energizing lifeless materials into living cells—wood on the inside, bark on the outside.

The fact that the heartwood is dead explains why a tree may be killed by making a circular cut around the outer bark and sapwood. It also explains why a tree may continue to grow and appear healthy even though much of its central core has been burned out by lightning or destroyed by decay. Such a tree has suffered no vital injury; it has merely lost some of its supporting structure.

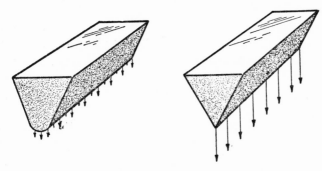

FIG. 20.   A SHARP AND DULL EDGE

A sharp tool cuts easily because the cutting force is concentrated in a small area. This produces a great pressure at the cutting edge.

Why does a sharp knife cut more easily than a dull one?

A sharp knife cuts through a material easily because of the great pressure it can exert. Although it is not scientifically correct, many of us commonly use the words *force* and *pressure* interchangeably. We speak of the number of grams of pressure that a needle exerts on a phonograph record when in reality, pressure cannot be measured in grams. We speak of a storm of great force when pressure is the significant characteristic. In a scientific sense, force may be defined as a push or pull. If we push against a desk or pull a heavy table into place we are using force. Force is commonly measured in such units as ounces, pounds, tons or grams. Pressure, on the other hand, is the force applied to a unit area of a surface. The weight of the atmosphere exerts a pressure of about 15 pounds on each square inch of surface it touches. The magnitude of this pressure is, therefore, 15 pounds per square inch. If a surface has an area of 1,000 square inches, the atmosphere exerts a force of 15 times 1,000 or 15,000 pounds on that surface. The purpose of sharpening a knife is to present as high a pressure (not force) as possible on the object to be cut. When a knife is dull, the cutting edge has a relatively large surface area. The force we exert, therefore, is spread out over this entire surface resulting in a comparatively low pressure. After sharpening, that same force is concentrated in a very small area of cutting edge resulting in a very high pressure. It is this increased pressure which results in easier cutting. In practice, a sharp knife may exert pressures of hundreds of pounds or more per square inch.

Can water boil and freeze at the same time?

It's only natural for us to associate boiling with heat because most ordinary things boil at high temperatures. Water, oil and cooking fats are typical examples. But some substances boil at very low temperatures. Liquid oxygen, nitrogen and hydrogen, all boil at temperatures below −180° C. Scientists tell us that the temperature at which a substance boils depends upon the nature of the substance, and the atmospheric pressure to which it's subjected.

When a liquid evaporates, particles escape from its surface into the surrounding air. Boiling, on the other hand, is a condition in which particles escape from *all* portions of the liquid, inside as well as the surface. Naturally, particles within the body of the liquid must form bubbles of gas to do this, and these bubbles are always present in boiling liquids. When a pan of water is placed on the stove, the water temperature rises until the boiling point is reached. Then the water begins boiling vigorously and the temperature will rise no further, no matter how hot the flame. For this reason, the rate of cooking can't be hastened by making the water boil more vigorously.

Water can be made to boil at temperatures from 0° to well above 100° C. by controlling the air pressure above the surface. The greater the pressure, the higher the boiling point. When the pressure is relatively high, water molecules escape only with difficulty. When the pressure is low, they encounter very little resistance in leaving the liquid. Imagine a flask of water, closed at the top except for a tube which leads to a vacuum pump. If we reduce the pressure in the flask about 200 times below normal atmospheric pressure, the water will boil at 0° C. If we reduce the pressure a bit more, some of the water will freeze while the remainder boils! When the pressure is about 200 times below atmospheric, water boils and freezes at the same time!

Do sea animals sleep in the water?

Yes, many of them do. The bottle-nosed dolphin, for example, sleeps in open water almost exclusively. During the night, the sleeping animal floats about a foot below the surface. About once every thirty seconds or so, a few gentle strokes of the tail bring its head to the surface for air. In a quiet tank, the sleeping dolphin appears to remain quite motionless. If a current exists, slow movements of the tail maintain its position constant with respect to the walls of the tank.

The sea otter is another mammal that prefers sleeping in the water. When the sea is calm, the otter selects a likely clump of seaweed, wraps itself up in it to prevent drifting, and goes to sleep on its back. The herd generally uses the same sleeping quarters during their stay in a given area.

The walrus sleeps while floating in a vertical position. The head is kept out of the water by the simple expedient of blowing up its neck! Perhaps one of the oddest of them all is an Antarctic animal known as Weddell's seal. This animal spurns the advantages of migration and spends the winter under the ice floe. It breathes by periodic visits to holes in the ice which it religiously keeps open by means of powerful teeth. How does it get any sleep? Perhaps the animals sleep in shifts. In any event, it's an intriguing question and nobody seems to have found the answer.

Can pictures be taken in the dark?

Light is one of the most fascinating mysteries of science. We see objects about us, and we distinguish colors by means of the light entering our eyes; but its exact nature remains obscure. We do know, however, that whatever light really is, we are able to see only a small portion of it. One of the characteristics of light is its wave length. Ordinary white light consists of a mixture of wave lengths from about 0.00008 centimeters (red) to about 0.00004 centimeters (violet). This range is called the *visible spectrum*. Light that is either above or below this range of wave lengths is invisible *to humans*. Infrared "light" exists in the region of wave lengths somewhat greater than 0.00008 centimeters. Invisible light of this kind is given off by all hot objects. By using photographic film that is sensitive to infrared light, it's possible to take pictures with no visible light present at all. All that is necessary for illumination is a hot body, such as an ordinary flatiron. Infrared rays leave the flatiron, are reflected by the scene to be photographed, and enter the camera to produce an image on infrared-sensitive film. Like ordinary light, it can be reflected, focused by the camera lens, and projected from one place to another.

What kind of senses do insects have?

Insect senses are quite different from our own. A bee can easily tell the difference between a sugar solution and a substitute such as saccharin, though both of these taste pretty much the same to us. We think of the sense of taste as being centered in the mouth, but a fly has taste buds located in his feet. All he has to do is walk around until he finds something that "tastes" good to his feet, and then eat it. You can test this ability by placing a few drops of sugar water on a table and watching a fly, or bee, as he wanders around. As soon as he puts his foot in it, he will get excited and start to drink. Amputate his feet and he will lose this ability.

The sense of smell is believed to be located chiefly in an insect's antennae. Cockroaches use their waving antennae as direction finders to locate the source of a particularly pleasant odor. In addition to the location of food, the sense of smell is sometimes used in locating a mate, or in determining the presence of natural enemies. Some of the gregarious insects use the sense of smell to follow one another in columns by tracing the scent of a predecessor. Of course, other insects are able to find their way around by sight.

Aquatic insects usually have a well-developed sense of balance. This results from pressure-sensitive receptors located on different parts of the body. When the pressures are the same, the insect is horizontal. Flying insects must have some sort of balancing mechanism too, since we never see them flying upside down. It is suspected that the fly's second pair of wings vibrate in such a way that they resemble the stabilizing action of a gyroscope.

Although insects have a hard skin which also serves as a skeleton, they appear to have a sense of touch just as we do. In many places, the hard plates are thinned out so that nerve endings can come close to the surface. In addition, there are small hairs which project from the surface and which are sensitive to touch and to the motion of air.

Hearing is quite variable in insects, but present in many species nevertheless. Caterpillars and crickets are known to be able to hear. Grasshoppers have the sense of hearing centered in their front knees. Other insects have their hearing located in the most unexpected parts of the body. The cicada's "ears" are located on the stomach, while the water bug has his in the thorax. Among the nonhearing insects is our useful friend the bee, who is as deaf as the proverbial fence post.

Seeing, in insects, is nothing at all like human sight. Their eyes, if any, are extremely simple and cannot possibly form images comparable in definition to our own. Some insects have light-sensitive elements under their skins. This must be similar to our ability to detect sunlight shining down on our bodies. The best eyes found in insects are called *compound eyes*. They are located on the side of the head and consist of a honeycomb of lenses. Each lens produces a picture on light-sensitive cells located beneath it. The optic nerve is connected to each of these cells and adds up the individual images to form one compound picture. It is as though each cell were a small frame of a stained-glass window. While any one frame is useless, added up they produce a somewhat intelligible picture. Compound eyes are more useful in detecting motion than in recognition of an object and many insects cannot recognize their prey until it moves.

Why does the Gulf Stream flow?

One of the odd features of the oceans is the existence of large masses of water which flow in much the same manner as rivers on land. Ocean-flowing streams of this kind are called *drifts*. As water near the Equator is heated, it is blown westward by the action of the *trade winds*. One such drift exists between Africa and South America. When the water reaches Brazil it divides, part flowing south and part flowing north toward Florida. Some of this water is heated further and flows north at four miles per hour as the *Gulf Stream*. When the Gulf Stream reaches the belt of the *prevailing westerlies*, it's deflected in an easterly direction toward the British Isles. Part of it flows past the British Isles, while the remainder diverts toward the south, flowing past Portugal and Spain as the *Canaries Current*. The Gulf Stream and Canaries Current are portions of the *North Atlantic Current*. It's this current that gives Western Europe its mild climate. It's quite a shock to glance at a map of the world and find that London is at the same latitude as Labrador! Were it not for the ocean drifts, the climate of these places would be more nearly the same. Land lying close to these currents may have a climate that varies as little as fifteen degrees throughout the year.

There are also movements of cold water throughout the oceans. One such current is the *Arctic Drift* which flows past Greenland, Labrador and Newfoundland. It then drops to the ocean floor to flow underneath the Gulf Stream. The Arctic Drift makes Labrador a cold and forbidding land in spite of the fact that its latitude is the same as that of England. Quebec and Newfoundland have a cold climate for the same reason, even though they are at the same latitude as France.

Are any of the mammals venomous?
Until recently, scientists believed that the use of venom was confined to reptiles and lower animals. Their ideas changed, however, when they became interested in the structure of certain cells of the salivary glands of the North American short-tailed shrew. While studying these animals, O. P. Pearson ground the submaxillary glands and mixed the material with a solution of salt. He then injected the liquid into mice and observed rapid paralysis, convulsions and death. A dose weighing as little as one seventy-thousandth of the mouse's weight was lethal in some instances. The gland of one shrew contains enough venom to kill over one hundred mice! It's surprising and somewhat disturbing to find this very reptile-like characteristic in a mammal. It would be interesting to know if any other animals possess this ability to kill with poison.

What are hormones?
In the advanced age we live in, new terms are forced upon us at every turn of the scientific road. Such a term is *hormone,* which covers a class of substances that influence the activity of our various body processes. Hormones are manufactured in the ductless glands of

our body. They eventually find their way into the blood stream and control many activities vital to our very existence. One of the many important hormones is *insulin*. It is produced by tiny groups of cells of the pancreas called the *islands of Langerhans*. Without insulin the body can't use sugar effectively and the patient comes down with diabetes. Insulin can now be extracted from animals, purified, and then given by injection to diabetics, enabling them to lead normal lives and eat normal foods.

Thyroxin, another important hormone, is produced by the thyroid gland. Scientists have determined its chemical formula and are able to manufacture it synthetically. Thyroxin stimulates the conversion of foods into energy. For this reason, many reducing medicines contain thyroxin to help "burn up" excess fat. There's a terrible danger in this, however, and thyroxin should be taken only under a physician's supervision. This hormone takes effect slowly but it remains active within the body for weeks. If an overdose is taken, body tissues may be consumed and serious illness or death may result.

Many hormones, all acting in concert, result in the development of a normal person. But if there happens to be a chemical imbalance of the hormones, people become ill, misshapen or defective. Most of the freaks in circuses are suffering from glandular imbalance. Many such afflictions could have been prevented, or can be cured by proper treatment. Even minor disturbances can be traced to hormone imbalance. Irritability, laziness and irresponsibility often originate with improper hormone balance. As strange as it may seem, we can get to understand others and ourselves better by merely learning more about hormones.

Which trees grow the tallest?
The loftiest of all the world's known trees is a redwood called the Founder's Tree. It towers 364 feet high in the Humboldt State National Park of California. Still growing, it has a circumference of forty-seven feet. If it is to maintain supremacy of the forest, it must keep on growing too, for another near-by redwood is only three feet shorter! Here, in a limited area, are located the world's tallest trees. Others, taller yet, would now be living in California had not early woodsmen reduced them to unexciting building materials. It is believed that the eucalyptus tree of Australia may once have

rivaled the redwood in height, but surviving giants of this type now average at least fifty feet shorter. Other runners-up include the Douglas fir and the Sequoia, specimens of which have grown well over three hundred feet tall. But in height, the redwood of California retains unchallenged supremacy of the forest.

Do flying squirrels really fly?

No, they don't fly, but they do glide. Our North American flying squirrel is an intrepid acrobat, gliding gracefully from the highest treetop to an adjacent limb or to the ground. He's suitably equipped for his aerial tendencies, having winglike membranes that stretch between the limbs and the sides. As he glides through the air, the squirrel controls his direction by varying the tension of his "wings" and the angle of his tail. He can turn sharply to avoid a collision and sometimes even seems to turn at right angles in mid-air. As he approaches the trunk that is to terminate his flight, the squirrel merely flips up his tail. This causes his body to turn upright, thereby checking his speed for the landing. His body comes to rest gently, and in the correct position for scampering up the tree preparatory to the next gliding descent. Except for the bats, who flap their wings in true flight, the squirrels are the most proficient aerialists among the mammals. Flying squirrels are also found in Asia, Africa and Australia. Some of these species are capable of "flying" for as much as a hundred yards without any apparent loss of height.

Do whales ever get the "bends"?

The "bends" are normally associated with human divers but is it possible for whales and other seagoing mammals to suffer from this disease? They take air into their lungs, as humans do; they descend to great depths, as divers do; and yet they seem immune to the effects of decompression sickness. The answer lies in several safety features built into these marine creatures by nature.

You will recall that a deep-sea diver breathes air that has been compressed in order to force it down into his suit. As a result of this pressure, the diver's blood dissolves more than the normal amount of nitrogen. If the ascent to the surface is too quick, the dissolved nitrogen forms bubbles which block blood vessels, causing severe pain. Whales don't suffer from this condition and a study of their breathing apparatus reveals why. In the first place, whales don't continue breathing while submerged. Their supply of nitrogen, therefore, is the same at the bottom of a dive as it was at the surface. It is not constantly renewed by an air hose as in the case of the human diver. And in addition, the increased pressure of the depths soon compresses the air cells of the lungs, forcing the contained air into the nasal passages and trachea. This further reduces the dissolving of nitrogen by the blood. Finally, the whale's heart begins to beat very fast on his return toward the surface. This helps to eliminate the nitrogen dissolved in the blood.

Before we leave the subject of whales, perhaps we ought to clear up the question of their lung capacity. The lungs of a man hold more air for his weight than do the lungs of a whale—in fact, twice as much! And in addition, the oxygen-carrying ability of the blood is about the same for both! Why then are whales able to stay submerged much longer than man? For one thing, whales exhale more completely than man. They renew 90 per cent of the air in their lungs with each breath compared to about 15-20 per cent in man. Moreover, whales are much less sensitive to carbon dioxide in their blood than man. It is the presence of carbon dioxide in our blood which acts on the nervous system to "force" us to breathe. These differences increase the amount of oxygen available at the beginning of the whale's dive, and reduce the stimulating effect of carbon dioxide on the animal's breathing. A third and equally important difference is the rate of the heartbeat. As soon as a whale dives, his heart slows down considerably. This permits more economical use of the oxygen that is stored up in the blood.

Why does popcorn pop?

We are all familiar with the miniature explosions that transform ordinary-appearing kernels of corn into popcorn. It was originally believed that the "pop" was caused by the expansion of air, or perhaps the changing of corn oil into gas within the kernel. But experts now believe that the popping of corn results from the rapid expansion of moisture within the kernel, and its sudden release when the wall of the kernel breaks open. There is a certain amount of water in the starch grains, and heat converts this water to steam. If the type of corn used has a particularly hard-shelled kernel, the steam pressure can build up to a significant value before any begins to leak out. When a fissure does appear in the kernel, the great pressure within throws out the pure white pulp of the interior.

The ability to pop seems to be an inherent characteristic of some varieties of corn. *Zea mays everta* is such a variety, consisting of small stalks, ears and kernels. Its kernels have harder walls than ordinary field corn, a characteristic that is essential if the kernel is to explode rather than slowly releasing the steam as it is built up. The pressure created by the conversion of water into steam must be contained with the grain until it has reached a relatively high value. Its sudden release then results in an explosion which carries the white starchy interior with it. The explosion is due only to the expansion of water vapor within the kernel, according to the experts, and not to the air or volatile oils contained in the corn.

Which insect is the most prolific?

Aphids, or plant lice, are voted by many to be one of the most prolific of the insects. In the fall, male and female aphids produce "winter eggs" that are hardy enough to endure the rigors of winter. When warm weather arrives, these eggs hatch into females which are capable of reproducing their young without the help of males. All of the young are born alive until the cold weather arrives again. At that time, males and females join in the sexual reproduction of winter eggs to complete the cycle. To make things even more complicated, some aphids even change the type of food they eat when they become sexual adults.

In an attempt to determine the rate of aphid reproduction, one investigator came up with an astounding conclusion. He states that, given a limitless food supply and freedom from natural enemies, one winter egg could produce enough offspring in one season to equal the weight of the earth! Fortunately for us, the aphid is eaten in great quantities by the lacewing flies and ladybird beetles.

If you would like to look at an aphid firsthand, they can be found in great numbers on the fast-growing shoots of rose bushes and many other plants. They obtain their food by sucking plant sap directly from the tender stems and leaves of the plant. Since the nutritive value of sap is relatively low, they must consume great quantities of it. This would normally require a large capacity for fluids on the part of the aphid's digestive system. To circumvent this difficulty, the aphid's digestive system is provided with a filter which quickly separates the food from the water. In this way the water is filtered off and released just about as fast as it is drunk while the food is concentrated and retained for the required digestive action.

What is the Milky Way?
Hundreds of years ago—before Galileo invented the telescope—men recognized several hazy patches of light in the nighttime sky. They called them *nebulae,* or clouds. The powerful telescopes in use today have revealed many more of these clouds. Astronomers tell us that they are really congregations of great numbers of stars as large as, or larger than, our sun. It's estimated that there are at least 75,000,000 star groups, or galaxies, in the sky. They range in size from very small families to the largest galaxy yet discovered, our own *Milky Way.* The Milky Way is a necklace of misty light that encircles a sector of the heavens. It consists of many millions of stars arranged in a form that resembles two saucers placed rim to rim with their bottoms out. The sun, with her family of planets, is located about two-thirds of the way from the center of the Milky Way to its edge. Virtually every star visible to the unaided eye is part of the Milky Way; but the greatest concentration is seen when we look in the edge directions of the "saucers." The galaxy next largest and nearest to the earth is the famous Great Spiral in the Andromeda. It's one of the few other galaxies visible to the naked eye.

Do electric eels really generate electricity?
The stories of electric eels and their ability to generate electricity were once considered greatly exaggerated. It's now known, however, that a large electric eel can generate hundreds of volts of electricity for a short time—quite enough to kill a man. Although they resemble eels, they are really a family of large South American freshwater fish related to the catfish, carps and suckers. Some of the larger species reach a weight of fifty pounds and a length of eight feet. They have been known to attack and stun horses that disturbed them while crossing a river. In addition to the electric eels, certain skates, rays and catfish also discharge electricity, but to a lesser extent. The generation of this electricity is connected with the nervous system of the fish. Scientists believe that all nerve impulses in animals are electric in nature. Minute electric currents are believed to provide the connecting link between the brain and the nerve endings of the body. It is believed that certain organs of

electric fish are abundantly supplied with nerve cells. These are probably developed to a much higher degree than in ordinary animals, enabling them to discharge much greater quantities of electricity. Experiments have shown that electric eels have a negative terminal near the tail and a positive one near the head. Some electric fish, such as the catfish of the Nile, have their terminals reversed. In all electric fish, however, the ability to give a shock is intermittent in nature. The "battery" soon runs out of power and a short rest is required to restore it.

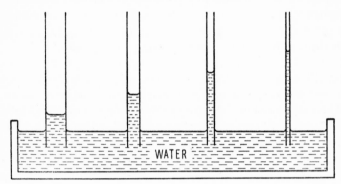

FIG. 21. CAPILLARITY

Capillary action causes water to rise in fine diameter tubes.

Why does plowing reduce evaporation from a field?

Most of us easily accept the notion that water seeks its own level. Yet in hundreds of ways, water seems to disobey this natural rule. Ink is absorbed by a blotter even though it must go uphill to do so. Oil rises in the wick of a lamp and water spreads through a towel with ease, regardless of the direction it must take. These and many other gravity-defying actions of liquids are due to *capillary action.*

If we place a very narrow tube in a glass of water, we find that water will rise in the tube. If we make the tube narrower, the water will rise even higher. This phenomenon is due to the combination of two actions: the tendency of water to wet or adhere to glass, and the attraction of water molecules for each other. At the instant we place the tube into water, the water climbs up the side of the tube a short distance due to its tendency to adhere to glass. The water molecules at the center of the tube then rise due to their attraction

186

for the molecules which have risen. The two processes are continued until the weight of the column of water in the tube just balances the tendency of the column to rise. This same action takes place with water in the soil. After a rain, water soaks into the narrow passages of the soil until it is well distributed throughout. As the surface water evaporates, capillary action causes other water molecules to rise and replace it. Evaporation takes place until the soil is completely dry. It is possible, however, to prevent this rise and escape by breaking up the surface soil through cultivation. This increases the size of the soil passages to a point where capillary action cannot take place.

What is the difference between trotting and pacing?
Most hoofed animals, as well as dogs and cats, walk on diagonally opposite feet. As an example, the right front foot and rear left foot might be on the ground at one time while the other two are in the air. The feet on the ground support and propel the animal forward while the other two get ready to replace them in the next step. Other animals, such as the giraffe, camel and brown bear, use a different type of locomotion, called pacing. In pacing, the two feet on the same side work as a unit. First the two left feet support and move the animal and then the right feet go down and take over. Trotting is similar to the diagonal step mentioned earlier, except for a period of suspension between the instants of support by one diagonal pair of feet and the other. The gallop is different from all of the previous types of locomotion. In the simplest case, the two rear feet push the animal forward. While in the air, the animal extends itself and lands on the front feet; it immediately shifts the hind feet forward and follows with another leap. Animals such as the ermine use this type of gallop in which the front feet and rear feet work in pairs. In many kinds of galloping animals, the feet hit the ground successively, the two hind feet usually making contact with the ground at more nearly the same instant than the two front feet.

Why do movies sometimes show wagon wheels going backward?
Ordinary automobile wheels photograph well in moving pictures, but wheels with spokes, such as are used on wagons and stagecoaches, have been known to drive more than one movie director to distraction. Spoked wheels insist on turning backward—forward—slowly

187

for a while—then faster, with complete disregard for actual conditions. This is a photographic illusion. It results from the fact that a moving picture is really a series of separate pictures shown in quick succession. A fraction of a second elapses between the showing of one picture and the next. Imagine that you're photographing a speeding stagecoach as it bounces across the prairie. Each click of the movie camera takes a separate picture of the scene. Suppose that the time between clicks is just long enough to allow a spoke to move to the position at which its predecessor was photographed on the previous click. In other words, suppose that each time a picture is taken, the rotating spokes happen to be located in the same angular position. Under these conditions, the wheels will seem to stand still even though the stagecoach is moving forward. Or worse still, suppose the spoke doesn't quite reach the same angle as that from which its predecessor was photographed. Each time a picture is taken, the spokes will seem to have backed off a trifle and the wheel will appear to move backward! What actually happens to the wheels in the movie depends upon the relative speed of the camera and the wheels. As the speed varies, the wheels seem to change from forward to reverse with annoying regularity.

Although this effect may be a nuisance in the movies, nature has seen fit to give it a useful side. A device, known as the *stroboscope,* makes use of this principle to accomplish a number of amazing things. It consists of a lamp that can be made to flash on and off at a rapid rate. In addition, this rate of flashing can be controlled by the operator from a few to many thousands of flashes each second. Through its use, engineers can make a rapidly moving machine appear to stand still! Here's how it's done. Before the machine is started, a white chalk mark is made on the moving part to be studied—a flywheel, for example. Then the machine is started and the room darkened. The stroboscope is turned on and the rate of flashing adjusted until the white chalk mark appears to stand still. This means that the time taken for one rotation of the flywheel is just equal to the time interval between flashes. Since the machine is in darkness between flashes, the white chalk mark and the entire flywheel look quite stationary. The illusion is so perfect that many workers have been injured by inadvertently touching such a "stationary" machine. When viewed under stroboscopic light, the operation of a machine can be studied in minute detail, enabling the designer to determine weaknesses and flaws that would otherwise go unnoticed. The principle that makes wheels in the movies go backward helps to turn the wheels of modern industry forward!

Do insects have slaves?

You will recall that the ancient Spartans all became soldiers, while domestic and industrial labor was performed by slaves. Close parallels to this kind of society exist in the insect world. Certain types of ants have lost the ability to reproduce workers. Colonies of such ants would consist only of males, queens and soldiers were it not for slavemaking. But these ants are incapable of feeding their young and cannot even obtain food supplies for the colony. Since only soldiers can be produced in each succeeding generation, such ants must obtain adequate numbers of slave laborers to do their work. The soldiers are usually well equipped for battling other ants and make periodic forays for slave pupae. Worker pupae are kidnaped from their nests and taken into slavery. When they mature, they labor for their masters as though they were among their own kind.

Hornets give us an example of a different kind of slave society. Certain kinds of hornets have lost the ability to produce any other than males and queens. In order to survive, such parasitic species must make use of the workers of another kind of hornet. The fertile parasite queen picks out a likely-looking nest, kills the reigning queen, and takes over the egg-laying activities. The adopted workers feed and nurse a generation of the parasitic queen's offspring. All of these are males and queens and they fly away upon reaching maturity. This leaves the adopted nest barren, and its whole social structure collapses.

When insects take to slavemaking, the slaveholders always seem to lose some important function. In the cases given above, it was the ability to produce workers to care for the young and perform the work of the society. It is not known whether this lost ability is a cause or an effect of slavemaking. The practice is highly developed, however, and necessary to the survival of the species involved.

Why do leaves change color in autumn?

Perhaps the most beautiful aspect of approaching winter is the turning of leaves in the forest, as the overwhelming green of summer gives way to the more delicate shades of autumn. Although this change is often ascribed to the action of frost, its real cause is more subtle and intriguing. The green of summer, of course, is due to the color of *chlorophyll,* the marvelous food laboratory of living plants. It constitutes roughly two-thirds of the leaves' pigmentation. Other colors are present too, however, even though their presence is eclipsed by the preponderance of chlorophyll. *Xanthophyll,* which consists of carbon, hydrogen and oxygen, is yellow. It makes up some 23 per cent of leaf pigment. *Carotin,* which gives the carrot its color, is also present in the amount of about 10 per cent. Carotin consists of carbon and hydrogen. Another pigment, *anthocyan,* gives the sugar maple and the scarlet oak their vivid red color. When cold weather approaches, the food products stored in the leaf begin to flow out to the woody fiber of branches and trunks. The food factory is closing down; the inventory is being reduced; the chlorophyll machinery begins to disintegrate. As the chlorophyll fades in the leaf, its fellow pigments begin to make their presence known in rich shades of yellow, orange and red.

The leaf is now dying and a thin layer of cells forms across the stem uniting leaf and twig. The formation of this *absciss layer* is sometimes accompanied by a softening material which aids in weakening the bond between leaf and plant. Eventually, the link becomes weak enough to give way to a stray breeze and the leaf flutters down to the ground.

Why do Thermos bottles keep liquids cold?

The Thermos or vacuum bottle consists of a silvered glass container, the double walls of which contain a partial vacuum. Heat escapes from any container by conduction through the walls and by radiation. Conduction takes place when a heated molecule, and therefore a fast-moving one, strikes a slower neighbor and imparts to it some of its greater mobility. In this way, heat (molecular motion) is passed from one molecule to another until it is lost to the outside air. By removing the air from the double-walled container, heat can be conducted out of it only at the neck where the two walls come together. Naturally, this part of the vacuum bottle is made as small as practical. Radiation of heat is minimized by silvering the inner and outer walls of the bottle. Highly polished, reflecting surfaces are poor radiators of heat. The combination of a partial vacuum and the silvered walls results in a container which retards the flow of heat out of, or into, the vessel. In this way, it's possible for the contents of such a bottle to remain at substantially the same temperature regardless of the temperature of the surrounding air. The first vacuum bottle was constructed by Sir James Dewar during the latter part of the nineteenth century. He needed such an insulating bottle for the storage of liquefied gases. Heat rapidly converts such liquids back into their gaseous form, whereas the use of his Dewar bulbs preserves them in the liquid state for relatively long periods of time.

What animal never drinks water?

It's difficult enough to believe that even one animal can live without drinking water, but there seem to be several of them right here on our American desert. Some pocket mice and a number of kangaroo rats manage very well on a diet of barley and oat seeds in which the water content does not exceed 5 to 10 per cent. And they never drink water! In scientific experiments, these rodents manage to live and grow fat on a diet consisting wholly of dried seeds. Under similar circumstances, ordinary mice lose weight and die. How do they do it? Do their body tissues build up a reserve of water upon which they can draw in the future? No. It's much simpler than that. They merely use water economically. Their urine is about twice as concentrated as that of other rodents. This results in such metabolic economies that the water obtained from food is adequate for their needs.

It seems that the giant kangaroo rat of California is even a bit fearful of getting *too much* water! During the later part of winter and early spring, this animal collects the seeds which are to carry him through a quiet summer of restful inactivity. But does the kangaroo rat carry them down to his subterranean storehouse? Not at all; he first distributes them in a series of small (an inch deep) holes in the dry earth around his home. He then covers them up carefully to make them invisible. When April comes around, the animal empties these caches and transfers the seeds to his underground home. Why does the kangaroo rat go through this two-step procedure of food collection and storage? Probably to let the seeds dry out!

What causes the holes in Swiss cheese?

The holes, or *eyes*, that we usually associate with Swiss cheese result from the action of bacteria during the process of fermentation. The bacteria, originally obtained from the stomachs of calves, produce gas within the cheese which collects to form holes, or eyes, sometimes as large as a half-inch in diameter. The presence of regularly formed holes with shiny walls is supposed to indicate cheese of good quality and flavor. It was originally believed that the environment of the Swiss Alps was essential for the development of eyes, but today excellent "Swiss" cheese is produced in the United States and elsewhere by employing a suitable bacterial starter. Steady improvement of the bacteria strains used has resulted in improved quality until, today, millions of pounds of natural Swiss cheese are produced annually in this coutnry.

How do bats navigate safely in complete darkness?

Bats are able to fly around in complete darkness without colliding with walls, branches or other obstacles. Although hard to believe, this was proved back in the eighteenth century by Spallanzani, the Italian naturalist, who blindfolded some bats and then found them perfectly able to fly around in complete safety. In 1920, the English physiologist Hartridge theorized that bats accomplish this feat by sending out supersonic sounds (very high in frequency, and inaudible to humans) that are echoed back to their ears. In 1940, two Americans, Griffin and Galambos, took up the study and established the soundness of this theory. It seems that bats give off bursts of supersonic sound in the range between 40,000 and 55,000 vibrations per second. This is just about two or three times as high in pitch as can be heard by human ears. Bats give off, on the average, between five and ten of these supersonic "cries" each second, increasing the rate to as high as sixty per second in the vicinity of an obstacle. These sound bursts are reflected by objects to be avoided and picked up by the bat's ears. The animal must subconsciously note the elapsed time between the origin and reception of the sound. Since sound travels at a definite rate, the bat gets a good idea of the obstacle's distance. If a bat's ears are plugged, in order to reduce his hearing efficiency, his skill in avoiding obstacles is reduced considerably.

You may wonder why many bats, all flying around in a group,

don't confuse their own supersonic "cries" with those of their companions. While it's not certain, it is possible that each emits a slightly different "sound" so that he can keep track of it and avoid confusion. In addition, these sounds travel only a short distance—perhaps five yards—so that the problem is not so great as it might be.

Perhaps the most amazing, and still unsolved, aspect of the problem is the bat's ability to distinguish between echoes from food insects and echoes from dangerous obstacles. Certain bats are able to whisk an insect off a branch or leaf as they dart past. How can these little mammals tell the difference between the two objects. If a small stone is thrown into the air below a bat, he will dive toward it and then swerve away from it at the last instant. The supersonic "radar" of bats, or *audiolocation* as it is known, is truly a magnificent and delicate mechanism.

What causes the northern lights?

The northern lights, or *aurora borealis,* are believed due to the impact of electrons from outer space upon rarefied particles of air in the stratosphere. Among the most beautiful and spectacular of natural phenomena, they have been seen at heights as great as six hundred miles above the earth. As seen in this hemisphere, their most common form is an arc of soft light in the north from which needle-like streamers radiate outward toward the sky. These streamers are never still for a moment. Sometimes they cover the entire sky with a mass of quivering flames, varying both in size and brightness. The color is usually a light green, but rose, lavender and violet tints are also common. They are most frequently seen when sunspot activity is at a maximum and the most brilliant displays occur when the largest sunspots are turned toward the earth. It seems probable that the aurora results from countless electrons arriving from the sun and colliding with the upper air of our atmosphere. The electrons are directed toward the poles by the action of the earth's magnetic field. Auroras in miniature have been created synthetically in the laboratory. Molecules of air in a glass tube are bombarded with electrons and caused to glow in wavering patterns similar to the aurora borealis. This tells us that the northern lights and their Antarctic counterparts, the southern lights, are quite similar to our ordinary neon lamps, differing mainly in size, color and regularity of the discharge. The southern lights owe their existence to the same principles and are known as *aurora australis*. The generic name for both the northern and southern lights is *aurora polaris*.

Which animal runs the fastest?
The swiftest animal appears to be the cheetah, or hunting leopard of India. A typical cheetah can reach a speed of 45 miles an hour in two seconds from a dead stop! After attaining full speed, a cheetah can outrun a horse or the fastest greyhound, reaching a top speed of over 70 miles per hour. There is a case on record of a cheetah covering over 700 yards in 20 seconds—a speed of more than 71 miles per hour. In comparison, the lion and gazelle can reach 50 miles per hour, the coyote and zebra 40, and the elephant 25. Camel and sheep lope along at about 10, while our nimble friend, the jackrabbit, can reach only 20 to 25 miles per hour. The fastest human can barely reach a speed of 20 miles per hour.

How do mariners use clocks to help determine their location?
The meridians of longitude are imaginary great circles drawn from pole to pole around the earth. By international agreement, the meridian of longitude passing through Greenwich, England is numbered zero. The earth is divided into 360 degrees and the meridians are numbered east and west from Greenwich. There are 180 degrees of longitude east of Greenwich and 180 degrees in the westerly direction. New York has a longitude of 74 degrees west (74° W.) which means that it lies on the 74th meridian west of Greenwich.

Since the sun appears to travel around the earth in 24 hours, it will move 360/24 or just 15 degrees in one hour. This reasoning can be used by navigators to determine their longitude. Imagine that we have set sail from Greenwich, England after having set a very accurate clock, or *chronometer,* to the local Greenwich time. As we travel westward toward New York, we notice that the sun is going "slower" than our chronometer. At the time that our timepiece reads 12 o'clock, the sun has not quite reached the zenith. As a matter of fact, when our clock reads noon, what it really means is that it's noon in Greenwich, England. Our clock continues to tell us the time, not at our present location, but at Greenwich. Let us wait until the sun is directly overhead (noon at our location) and then read the time on our clock. Suppose it reads 1 o'clock. This means that there is one hour's difference in time between our longitude and that of Greenwich. As we mentioned earlier, this corresponds to exactly 15 degrees of longitude, so our longitude must be 15° W. The world is divided into 24 time zones and each zone corresponds to 15 degrees of longitude. New York is approximately 5 time zones west of Greenwich so the time difference must be about 5 hours. By maintaining chronometers on Greenwich time, ships can determine their longitude on any sunny day by merely noting the difference in hours between Greenwich time and local sun time and multiplying this difference by 15 degrees.

Of course, longitude gives only half of the information needed to determine our precise location. We must also know our latitude, which tells us how far we are north or south of the Equator. The Equator is the zero line for the measurement of latitude. Circles are drawn parallel to the Equator to indicate other values of latitude. There are 90 degrees of north latitude and 90 degrees of south latitude. In the Northern Hemisphere, there is a star called Polaris almost directly over the North Pole. This makes it possible to determine the latitude of a given point by measuring the angle between Polaris, the North Star, and the horizon. Mathematicians tell us that this angle is equal to the latitude at the point in question.

To get an idea of our location, therefore, we need to know local time, Greenwich time, and the angle between Polaris and the horizon.

Are Japanese beetles a problem in Japan?

The Japanese beetle, like many other alien insects, presents no particular problem in its homeland. And it would present no problem in this country were it not for the almost complete absence of its natural enemies. In Japan the beetle behaves in much the same manner as its harmless American cousins do here. Its numbers are limited by a variety of natural enemies which regard the beetle as a rare delicacy. Suddenly, the insect was transported to this country, where there was plenty of food and no parasitic or predatory insects and birds that seemed to disturb it. Each female Japanese beetle lays about fifty eggs in a lifetime. In Japan, only about two of these reach maturity, one each to replace the mother and father. This keeps the population about the same from year to year. In America, almost all of the eggs develop into mature beetles. At this rate, one female can account for almost 2,000,000,000,000,000 beetles in the tenth generation, a fact that helps to account for the phenomenal increase in numbers and importance of this insect pest.

How are color photographs made?

An early method of making color photographs was based on taking three individual pictures, one for each of the primary colors of light: red, green and blue-violet. Light entering the camera was divided into three parts. One part was passed through a filter that would allow only red light to get through. This light was projected onto a film which formed an image according to the red light in the original scene. The same thing was done for the other two colors, resulting in three exposed films; one produced by each of the three primary colors. The films were then developed and dyed to correspond to the color of light they represented. When the three transparencies were projected onto a screen, the separate colors added together to duplicate the original colors photographed. The method in use today uses a film having three separate layers, each sensitive to light of a different color. By means of a rather complicated development procedure, each of the three layers is dyed with the appropriate color. These are then properly arranged to reproduce the colors of the original scene when light passes through them.

How high can a fish jump?

All of us have heard of the supposed ability of fish to jump up high waterfalls. It just isn't true. The best jump a fish can make is perhaps five or six feet—about the same as a man. But isn't it true that salmon often negotiate waterfalls that are twenty feet high? Yes, but they don't jump; they swim. Many fish, such as the salmon, are able to move very well against fast currents. Upon approaching a steep waterfall, they get a "running start," leap up part of the way, and swim up the rest. How about Niagara Falls? Don't fish manage to swim up to the top of this great cascade? No, they don't. Some fish can crawl across great stretches of dew-covered pastures (eels), and some fish can climb trees (oriental perches), but swim up Niagara Falls? Never.

Who invented the clock?

Once more we are indebted to the genius of Galileo. While attending Mass in the cathedral at Pisa, Galileo's attention wandered to a chandelier which was left swinging slowly back and forth after it was lighted. Since he had no timepiece, he timed the length of each oscillation by counting his pulse. He noticed that the time required for one complete oscillation was the same, whether the chandelier was swinging in a wide arc or a small one. This aroused his interest and he began to study the pendulum more closely. He found that for a simple pendulum, consisting of a weight and a very light string, the time required for one oscillation depends only upon the length of the string. Since the time required for one oscillation is invariable, Galileo reasoned that the pendulum could be used to measure time.

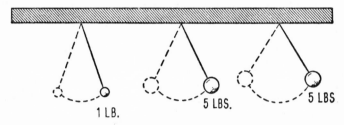

FIG. 22.   THE SIMPLE PENDULUM
All of these pendulums have the same length and, therefore, take
the same length of time for a complete swing.

Although blind, and imprisoned for his heretical writings, he later dictated to his son the procedure to be followed in the construction of a pendulum clock.

How do boa constrictors kill?
The common belief that constrictors mangle and crush their victims to death is nothing more than vivid fancy. Actually, a constrictor is quite subtle in the use of his great strength. After selecting his prey, the snake throws several quick coils around the victim and takes up all of the slack. Sooner or later the victim must breathe, and in so doing, exhales air. The snake then merely tightens its grip further, repeating the process whenever the captive moves or breathes. Circulation and breathing soon stop and the animal dies. The constrictor's sensitivity to the instant of death is another of its unusual characteristics. As soon as the coils fail to feel the pulse of life in the victim, the snake's job is finished and it relaxes.

197

Why are oil and gasoline tanks painted a silvery color?
Some day when the sun is shining brightly, touch the black fender of a car and then touch the chromium-plated trim. You will find that the dark parts are quite hot while the light-colored, reflecting parts are cooler. Black surfaces get hotter than white ones because they absorb most of the sun's radiation and reflect little. This absorption of light energy from the sun causes an increased movement of molecules in the dark object, and it gets warm. The easiest way to prevent things from heating up in this way is to paint them white or silver. This helps them to reflect most of the sun's rays. Many oil and gasoline storage tanks are painted a silvery color which reflects most of the sun's light and infrared (heat) rays. This keeps their temperature lower and reduces the risk of fire and explosion.

Thus far we have discussed the absorption of light rays and their conversion into heat. Now let's look into the influence of color upon the emission or radiation of heat from a hot body. Scientists tell us that light-colored objects are *poor radiators* of heat. In line with this principle, many modern homes are built with light-colored roofs. They reflect the sun's rays in the summertime, thereby keeping the home cool, and they radiate little heat in the wintertime, thereby helping to keep the home warm! A black roof has just the opposite effect. It heats the home during the summer and helps to cool it in the winter.

Why do lobster hatcheries have sliding chutes?
Although lobsters are prolific, and live in almost inaccessible locations, they would probably long since have become extinct were it not for the establishment of lobster hatcheries along our North Atlantic coast. Every female lobster produces thousands of eggs each season but only a small percentage of these manage to live to maturity. Of those that succeed, a high percentage find their way into the lobster pot and eventually the dinner table of seafood lovers. In time, lobster hatcheries were established to keep the supply of the delectable crustaceans in balance with the demand. All did not go too well at first, however, for the addition of millions of the youngsters to the lobster grounds appeared to have no significant effect on the adult population. Being delicious morsels, the youngsters were gobbled up by all kinds of carnivorous sea creatures nearly as fast as they were released. Their tender bodies and the lack of

powerful adult claws made them easy prey for their predatory enemies. But these same deficiencies afflict their wild cousins—and were it not for lobster fishing the creatures would soon multiply to their former great numbers. What could be the difference between wild and hatchery-raised lobsters? Quite by accident, someone discovered that artificially reared youngsters did not dive to the bottom and hide among the rocks and crannies as did their wild cousins. Raised in the safety of their hatchery environment, they had never learned to hide. When released, they merely swam around near the surface enjoying the luxury of their new and spacious quarters. This made them easy prey for fish and other crustaceans that took advantage of their inexperience to enjoy a delicious meal. It was clear that young hatchery lobsters would have to be taught to find suitable hiding places if the adult population were ever to be increased. This was accomplished by sliding the young lobsters down a chute to the bottom of the tank. This process was repeated over and over again until the young fellows learned to dive of their own accord. When released in the ocean they promptly dove to the bottom, just as wild lobsters do. Today, lobster hatcheries teach the youngsters how to dive, and the adult population is increasing. Unless the demand gets much greater, or more efficient catching methods are devised, there is little danger of fishing the creatures out.

Will the North Pole always be cold?

Most of us have heard of the four ice ages and regard them as interesting but relatively unimportant happenings of the past. I think you will agree that it comes as somewhat of a shock to discover that we are presently living in the tail end of the last one! According to scientists, we seem to be passing through the closing phases of the *Pleistocene Ice Age*. At its height, this ice age produced enormous sheets of ice that spread out from the poles until they covered most of the earth's surface. At times, the ice blankets were thousands of feet deep! When this ice age is completely over, and the earth gets as warm as it's going to, the Arctic regions will be warm enough for comfortable living. It's also possible that the North Pole will be warm enough to support the kind of plants and animals now found in Canada and other northern countries. Unfortunately, the warming of the polar regions will entail a few sacrifices. The melting of enormous quantities of ice will add materially to the supply of water in the oceans. This will increase the ocean depth by about a hundred feet, inundating the lower portions of many coastal cities. Luckily, this won't occur for perhaps another twenty-five to fifty thousand years. It seems likely that our present climate will remain substantially the same for many hundreds of years.

How can termites digest wood?
There are few substances less digestible than wood and yet termites seem to thrive on it. They owe their ability to digest wood to one of the most unexpected partnerships in the animal kingdom. All termites are infected with a quantity of single-celled, microscopic animals called protozoans. Protozoans inhabit the intestines of the termites, and obtain their food supply directly from the termites. In exchange, the protozoans secrete a substance which aids the termite in the digestive process. Without this substance termites die of malnutrition. This dependence of the termite on intestinal protozoans affords an unusual means of termite control. The protozoans can be killed at a lower temperature than the termites. This makes it possible to increase the temperature sufficiently to kill the protozoans, while not affecting the termites. In a few days, however, the termites starve to death. One of the oddities of this termite-protozoan relationship is the complete interdependence of the two organisms. Termites cannot live without their protozoans and if the protozoans are removed from the termites, the protozoans soon die off. Each creature is clearly dependent on the other. Separation results in the death of both. Partnerships of this sort where each organism depends upon the presence of the other are called *symbiosis*. There are over 160 known species of protozoans living within termites.

Why do some materials conduct heat better than others?
Imagine an aluminum pan resting on one hand while you pour boiling water into it with the other. In a short time, the pan becomes too hot to hold. If the pan is made of copper, it takes even less time for you to feel the heat. If it's made of iron it takes longer. If it's made of Pyrex glassware, the time required is much longer still. Such everyday experiences illustrate two important facts about heat: heat is conducted through solid materials; and some materials

are better conductors of heat than others. We make use of these facts every day by selecting the proper material for the problem at hand. We make cooking utensils of good conductors, and their handles of poor conductors. We protect our homes against cold in winter and heat in summer by putting poor heat-conducting *insulators* between the inner and outer walls. We cover steam pipes with asbestos in order to keep the heat in the pipe, and the surroundings cool.

But what is heat, and why does it pass through solid objects? Heat is caused by the motion of molecules, the building blocks of which all materials are made. Suppose we heat the bottom of a pan over the kitchen stove. The molecules closest to the flame get hot and begin to vibrate violently. These rapidly moving molecules collide with the slower-moving ones next to them, causing the latter to move more rapidly. These, in turn, strike others until the entire pan is heated. Solids are the best conductors of heat because their molecules are most closely packed together. This increases the occurrence of collisions and makes the transfer of heat more rapid. Gases are the poorest conductors of heat because their molecules are so much farther apart. Liquids lie somewhere between the two other groups both in molecular separation and in their ability to conduct heat. In comparing the heat conductivity of various substances, air is usually used as the standard. On this basis, silver conducts heat about 19,300 times better than an equal volume of air. The relative conductivity of a few common substances are given below.

Material	Relative Conductivity
air	1.0
rock wool	1.8
cork	1.9
linen	3.8
asbestos	4.8
wood	7.0
water	25
concrete	35
glass	44
iron	2,900
aluminum	8,800
copper	17,400
silver	19,300

How do starfish open oysters?

It's not at all uncommon to find starfish crawling around among the seaweed and rocks of our East Coast tidal pools. They look harmless enough as they meander about, but they are actually the relentless foe of oysters. Just like man, the starfish has a voracious appetite for oysters on the half-shell. But, you might ask, how in the world does a little starfish manage to open an oyster? Any amateur who has tried knows that it is no mean trick, even with proper tools. But what the starfish lacks in strength and natural ability, he makes up for in perseverance and tenacity of purpose. When a starfish is hungry he selects an oyster and proceeds to wrap his arms around it. Thousands of sucker-tipped tentacles adhere to the shells and the starfish pulls steadily upon the two halves of the oyster. Strong as an oyster's muscles are, they eventually get tired and the shells begin to part. When this happens, the starfish squirts some digestive juice between the shells and the oyster's doom is sealed. Groggy from this juice, the oyster opens up and the starfish has won the battle. The starfish, who is always hungry, then goes off looking for another oyster.

Another unusual characteristic of starfish is their ability to grow new arms, or rays, when one gets cut off. They can even grow a new body from a single amputated arm! If you want to reduce the starfish population, don't do as some Chesapeake Bay oystermen did, and cut them in half, returning the pieces to the water. Such a procedure merely results in two starfish where there had been only one.

Why do helicopters sometimes have small propellers on their tails?

If no precautions are taken, a helicopter tends to spin in a direction opposite to that of the whirling overhead blades. How does this happen? It's all a result of Newton's third law of motion—"To every action there is an equal and opposite reaction." When you jump upward, two things happen: *you* move up from the earth, and the *earth* moves "down" and away from you. It may be a small amount—an immeasurably small amount—but it moves nevertheless. When you step out of a boat, the boat moves away from you as you struggle for the pier; again two motions in opposition to each other. The water gushing forth from a fire hose exerts an opposite force on the fireman tending to push him backward; sometimes several men are needed to hold it steady. All of these are ex-

mples of Newton's third law. Every force is accompanied by an
equal and opposite one. Every acceleration (or change in motion)
s brought about because the moving object pushes back against
omething else. In a helicopter, this problem can be very serious
ince the force turning the propeller has an equal and opposite reaction tending to turn the helicopter in the opposite direction. Two
sets of overhead blades are often used to overcome this effect; each
rotates in a direction opposite to the other. If only one set of blades
is used, a small propeller on the tail controls the tendency of the
plane to rotate. This keeps the helicopter from spinning around in
circles.

Do insects get sick?

While we don't often stop to think about it, entomologists tell us
that insects suffer from the same kinds of diseases that we do. They
contract virus infections, bacterial plagues, fungus diseases, worms,
and they are subject to parasitism by other insects. The first recorded case of attempted medical aid to insects was connected with a
protozoan disease that threatened to wipe out the silkworm industry
in France. Louis Pasteur was called in to see what could be done.
While he didn't effect a cure, he did find out that certain individual
silkworms were immune to the disease. He bred and interbred these
worms until he had developed a whole immune population. His
work is acknowledged to be the first scientific study of an insect
disease.

Honeybees are normally quite hardy, but they too are subject to
a disease which accounts for practically all of the losses in honey
and beeswax production. It is a bacterial plague called *American
foul brood*. In many places, the authorities examine hives in order
to destroy infected bees and prevent the spread of this highly infectious disease.

Certain insect diseases, such as the "Milky Disease" of the Japanese beetle, are extremely useful to man. The bacterium, *Bacillus
popilliae*, is the cause of this disease. It operates in the ground on
the larvae of the Japanese beetle, causing them to contract the
disease and die. During the illness, the disease bacteria multiply
enormously and thereby make the ground even more infectious for
the next generation of beetles. Most of the Japanese beetle larvae
are killed after the proper distribution of the bacteria in the ground.

The common housefly is another insect that can get very sick.
About 50 per cent of the houseflies infected with the germ, *Staphylo coccus muscae,* are killed by it. They are also subject to a fungus disease which is similar to ringworm and athlete's foot in humans. The disease, caused by the *Empusa muscae* fungus, attacks and kills millions of flies every fall. A similar fungus acts in the same way on grasshoppers. In both cases, the fungus uses the insect tissue for food, eventually causing death.

In addition to the diseases mentioned above, insects suffer from a great variety of viruses such as *grasserie,* the jaundice of silkworms. Cholera and malaria-type organisms also infect certain insects. The Anopheles mosquitoes, which carry the malaria organism, contract "ulcers" from it for their pains. We might consider this to be a sort of poetic justice.

What is the oldest living thing on earth?
According to Mr. C. K. Bennett, a California naturalist, the oldest living thing may well be a juniper tree (*Juniperus occidentalis*) which he believes to be 6,000 years old. Named the Bennett Tree by the National Forest Service, the juniper stands 87 feet high and measures over 57 feet in circumference. If it is as old as it seems to be, it must have sprouted concurrently with the ancient Babylonian Empire. It would have been a solid thousand years old by the time the Egyptians were learning how to construct the first pyramid.

The tree stands nearly two miles above sea level on a ridge in California's High Sierra. Like the rest of its kind, it thrives on soil of volcanic origin amid the rocks and rugged terrain of mountainous regions. Tremendous root systems hold these trees firm against battering winds and heavy snows.

Mr. Bennett arrived at the estimate for the tree's age by borings made into several sides of the tree. Such measurements indicate that it took nearly 1,000 years to add the last foot of growth to the tree's radius. A limb, 3½ inches thick that had fallen from the tree was examined and found to have 550 rings. This showed the small limb to be 550 years old. Even if the tree is not as old as Mr. Bennett believes, it is undoubtedly one of the oldest living things on earth.

No discussion of this kind would be complete, of course, without mention of the giants of Sequoia National Park. The General Sherman has a girth of 101 feet 6 inches and its colossal trunk towers

272.4 feet above the ground. Its age is variously estimated between 3,000 and 4,000 years in spite of its exuberant vitality. If this giant Sequoia is indeed 4,000 years old, it seems quite capable of living as many more.

Are sponges animal or vegetable?

Most of us know that sponges come from the sea, but the fact that they are the skeletal remains of living marine animals probably comes as somewhat of a surprise. Even more surprising is the physical appearance of the living creatures, for they bear little resemblance to the sponges we see in daily use. Even the finest, softest varieties are rough, hard, slimy things when alive. They range in color from brown or black, to yellow, white, red and even vivid scarlet. There are blue sponges, green sponges, orange sponges and even variegated sponges. Some are round, while others resemble vases, human hands and a multitude of other everyday shapes. They vary in size from the very tiny, no larger than a dot on this page, to the very large, measuring several feet in each dimension. They may be found in all parts of the world from tropical tidal pools to Arctic seas. While some are of commercial value to man, others are his enemies. Certain sponges bore holes in oysters and even in stone and concrete. Some form huge colonies over oyster beds and smother the delicious bivalves. Still others destroy seawalls, jetties and other man-made structures.

All sponges are alike, however, in that they consist of many individual one-celled animals, joined together by a framework of minute, needle-like, bony material. Sponges reproduce by means of eggs, and free-swimming young that attach themselves to a shell or piece of coral where they settle down to routine living. Each such animal then begins to divide itself into parts which are exactly like the original. This process is called budding, or cell division. It enables the sponge to grow quite rapidly since a single cell may divide itself into ten cells, after which each of these divides into ten more until a colony of hundreds of cells may result. All of these cells are connected by the bony framework mentioned above. Water and food reach the internal sponge cells by means of a network of chambers, tunnels and canals within the sponge. If the sponge is cut into pieces, each piece will begin to grow just as if it had started in the normal way. This peculiarity has made it possible to cultivate desirable varieties of sponges on a large scale. Clippings of live sponges are placed on submerged concrete slabs and allowed to grow. Their rapid growth soon provides a great quantity of high-grade, symmetrical specimens. Moreover, when the crop is harvested, a small piece of each animal remains attached to the slab, forming the nucleus from which the next crop will grow.

After the sponges are harvested, they are placed in enclosed areas of shoal water. The animal matter decays from the bony framework, and then the sponges are cleaned, beaten and rinsed. When the sun has dried them out they are ready for market. During this period of preparation, great care must be taken to protect them from contact with rain or any other form of fresh water. Such contact rots the sponge material, producing rust- or orange-colored spots on a yellowish surface. If such sponges are offered to you for sale, resist their attractive appearance and select the really valuable dull buff or brown sponges that will give the best service.

Are the continents moving?
A German scientist, Alfred Wegener, was examining some maps back in 1910 when he was struck with an intriguing idea. The coastal contours of Africa and South America suggested that these two continents were once attached *and had drifted apart*. "He who examines the opposite coasts of the South Atlantic Ocean," he wrote, "must be somewhat surprised by the similarity of the shapes of the coast-lines of Brazil and Africa. Every projection on the Brazilian side corresponds to a similarly shaped indentation on the African side." It is as though two pieces of a jigsaw puzzle had been moved apart. These coastlines complement each other and can be fitted together to make a complete whole.

In addition to the coastline similarity, Professor Wegener soon learned that naturalists were discussing the similarities in the pre-historic plant and animal life of South America and Africa. This confirmed his idea and he then formulated his *theory of the displacement of continents*. Stated briefly, his theory assumes that the land masses of the earth were once joined together in a continuous, united continent. Just as at present, the land area was dotted here and there with a river, a lake or an inland sea. Eventually, for reasons unknown, the land mass began to break up. South America split off from Africa and floated to the west. North America left Western Europe in the same way. All of the continental land masses, as we now know them, were formed by this process of displacement.

Of course, all of this is theory. It's by no means an accepted principle. We do know, however, that the earth's crust is continually shifting. Our Pacific coast, for example, is rising at the present time while our Atlantic coast is sinking. This turning process is very slow, however, and cities like Baltimore and Philadelphia will remain above the sea for many hundreds of years. If you're planning on a long-term investment in real estate, however, swampland on the Pacific coast holds more promise than similar land on the Atlantic.

What are the White Cliffs of Dover made of?

Roughly a billion years ago the greater part of the earth's surface was covered with water. Microscopic animals, the protozoa, inhabited those prehistoric oceans in untold numbers. Although they are the oldest of all known animals, the protozoa of today are little changed from their ancestors of that prehistoric age. Some are soft-bodied creatures; some, called the *Foraminifera*, have hard lime shells; but all are alike in that they consist of only one cell. Protozoa multiply, not by laying eggs, but by constant subdivision of the adult into two new individuals. Although most are microscopic in size, a few are large enough to be seen with the unaided eye. Protozoa inhabit sea water in almost inconceivable numbers. They swarm in such great quantities that their skeletons have formed a mudlike deposit on the ocean's floor thousands of feet thick, covering hundreds of thousands of square miles. A single ounce of this material contains more than three millions of their shells.

Since the early days of the protozoa, the earth's surface has undergone many violent changes. Places that were once on the ocean floor were raised high above the water's surface to form land. In some places, the skeletal remains of the Foraminifera were changed into white-chalk formations such as the White Cliffs of Dover. In other situations of pressure and age the upraised ooze was changed into marble. Viewing a piece of marble under a microscope will reveal the remains of the minute protozoans of which it is composed. It's hard to believe that our finest statues, the famous white steps of Baltimore, and even the great pyramids of Egypt are composed of the microscopic shells of sea creatures of an unrecorded age. The fossil remains of protozoans provide us with chalk, marble, abrasives, polishing materials and certain medicines. In addition, these abundant creatures provide the principal food of many fish, lobsters, crabs, shrimp, corals, sponges and even some whales!

In the course of history, many creatures have appeared on the earth, or beneath its waters, only to remain for a while and then disappear. Of the forms that have survived, many have changed radically in an attempt to remain competitive in a highly changeable environment. But the lowly protozoan, simplest and most abundant of all living creatures, has managed to remain practically unchanged and certainly undiminished in numbers since the dawn of life on earth. There can be little doubt that the protozoan is eminently well suited to life in a world where only the fit seem to survive.

How does a hydraulic press exert such high pressures?

During the early part of the seventeenth century Blaise Pascal discovered that a pressure applied to a confined liquid or gas was transmitted equally in all directions. Using this principle, he constructed the first hydraulic machine, which consisted of two cylinders connected at their bottoms by a watertight chamber. One cylinder had a diameter a hundred times as great as the other and into each cylinder was placed a tight-fitting piston. Pascal discovered that a force of one pound pushing against the smaller piston could balance one hundred pounds pushing against the larger one. Pascal's device was actually a machine which could multiply force. He had discovered the hydraulic equivalent of the lever and wedge. After further investigation he found that pushing the smaller piston a distance of one inch would raise the larger piston only one-hundredth of an inch. By combining these observations he concluded that the multiplication of force was obtained at the expense of the distance through which it moves. As in all mechanical devices, the acting force times the distance through which it moves must be equal to the resisting force times the distance through which *it* moves. The hydraulic press operates on exactly this principle with the addition of a lever to operate the smaller piston. Refinements, such as valves, can be added to prevent the liquid from returning to the chamber after each operation of the smaller cylinder. This enables the larger piston to be raised higher and higher with each successive operation of the small piston. Such presses are used for baling cotton, squeezing apples, punching holes in steel plates. forming automobile bodies and operating barber chairs.

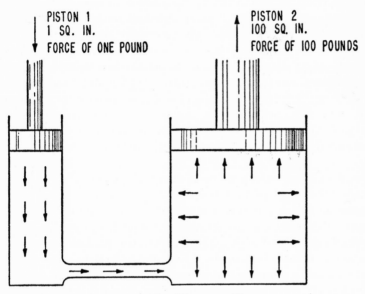

PISTON 1
1 SQ. IN.
FORCE OF ONE POUND

PISTON 2
100 SQ. IN.
FORCE OF 100 POUNDS

FIG. 23. THE HYDRAULIC PRESS

A small force exerted on the small cylinder produces a great force on the large cylinder. The pressure is transmitted from one cylinder to the other by a confined liquid such as oil or water.

Are sea worms edible?

Most of us enjoy oysters on the half-shell, or clams, crabs, shrimp or lobsters without much thought to their edibility. What we consider edible is really a matter of custom. Once custom and habit have given their approval, we concern ourselves only with the gastronomic qualities of a food. To most of us, crustaceans and bivalves are delicacies while worms are too repugnant to be considered seriously as food. But this is not so in the South Pacific islands. In these warm waters, a sea worm, called the *palolo,* is regarded as a highly prized delicacy. This in itself is unusual enough, but even more remarkable is the definite and amazing sense of timing shown by this worm.

209

The palolo worms wisely spend the greater part of the year hiding in holes and crevices in the coral reefs and formations around the islands. This provides them with a good source of food while protecting them from their natural enemies, which include the palolo-loving natives. But on the day before the moon is in its last quarter, in the months of October and November of each year, the palolos swim to the surface by the millions. Actually, only half of each worm makes the trip—the back half containing the eggs. The front end, with head attached, remains in its coral home. Palolos vary in color from deep green through blue, scarlet and yellow. Their exodus transforms the sea into a fantastic splash of color of every conceivable hue. The natives, however, have more than an aesthetic interest in the little creatures. Counting on the palolos' accurate sense of timing, they fill their canoes with baskets and set out to sea to await the great swarm. As soon as the worms appear, they are scooped out of the water in tremendous quantities and rushed ashore, for they don't keep too well out of water. Once ashore, the natives gorge themselves with raw or cooked palolos depending upon individual preference. While we may squirm a bit at the prospect of eating worms, whether of the sea variety or not, many Europeans living on these islands join with the natives in attesting to their delicious taste. When we consider the situation rationally, we are almost tempted to try a palolo—almost, but not quite!

Is there life on Mars?

Mars is the next planet beyond the earth and many regard it as the twin of our own planet. It has about half the diameter of the earth, rotates around the sun in somewhat under two years, and has a day that is almost equal in length to our own. It has seasons like the earth, an atmosphere containing at least a small amount of water vapor, and two small moons, Deimos and Phobos. Mars shows immense white polar caps which astronomers believe to be thin coverings of frost rather than ice, because of the scarcity of water on the planet.

In 1887, Giovanni Schiaparelli, an Italian astronomer, announced the discovery of long, straight and narrow (about fifty miles wide) streaks on the surface of Mars. Although some observers believe them to be optical illusions, it seems probable that the markings do really exist. This has naturally raised speculation concerning the possibility of living creatures on Mars that might have constructed these geometrical patterns, perhaps as canals to carry precious water to the warmer areas of the planet. The temperature of the planet, 80° F. at the Equator, is warm enough to support some form of life. It's quite possible that some forms of Martian life do exist. If they do, however, it would be impossible to speculate on their true nature. The atmospheric pressure on Mars must be much less than ours on earth due to the extremely thin atmosphere. The planet has little if any oxygen in its atmosphere. The water supply must be very scarce. Taking these and other facts into consideration, scientists believe that Martian life, if it exists, must be very different from any of the thousands of forms that we are familiar with on earth.

How long do ants live?
An ant colony consists of the queen, males, workers and soldiers. The life span of these types is as varied as their social functions. Males live only a short time, as their function is simply to provide fertilization to each succeeding generation of queens. Workers and soldiers live much longer, and the queen longer still. The workers may live as long as six or seven years in the average colony. A queen may still be fertile when she's ten. The queen's longevity is a good thing for the ants, for when she dies, the entire social organization of the colony collapses. They seem to realize her importance for they keep a dead queen's body around until little is left of it. Finally, however, the colony distintegrates since there are no new workers and soldiers to replace those that die or are lost.

What are cosmic rays?
The discovery of radium aroused in scientists a great curiosity about radiation and they began to look for it everywhere: in water, in air, in soil. To their amazement, they succeeded in finding it in just about every spot searched. It seemed to be the strongest near the earth and to diminish with altitude. It was expected that all evidence of radiation would disappear at some elevated point in the atmosphere. In order to find out, Albert Gockel, a German physicist, went up in a balloon to measure radiation at the higher altitudes. To his surprise, he found that it decreased with altitude to a certain point, and then started to increase again as he rose higher! He

theorized that an unknown radiation bombards the earth from cosmic space! Although he had made one of the greatest discoveries of the twentieth century, his contemporaries of the early nineteen-hundreds called him a fake, and his theory nonsense. After his death, however, one of his pupils succeeded in verifying his observations and Gockel was posthumously awarded the credit for his celebrated discovery.

What happens is this: The earth is constantly bombarded with a barrage of atomic particles. They are believed to be mostly protons, with a few larger particles, consisting of a combination of neutrons and protons, thrown in. They arrive at a speed of about three-fourths that of light. This tremendous speed gives each particle a fantastic amount of energy. They strike atoms of air, breaking them up into subatomic particles and energy that is given off in the form of "cosmic rays." The process is repeated about fifteen times as atom smashes atom until a shower of radiation rains down upon the earth. One proton from outer space accounts for a downpour covering an area of about a thousand square feet. This radiation consists of about 25 per cent mesons, another subatomic particle. Thousands of these particles pass through your body each minute on their way deep into the earth. Are they harmless? Apparently so, for life has existed for millions of years in the presence of their unending bombardment.

The origin of cosmic radiation is somewhat more obscure than its nature. We know of no source that could reasonably be expected to propel missiles through space at such enormous speed. The sun and stars are much too cold. We do know, however, that cosmic rays constitute our greatest potential source of energy. They dwarf all of our coal and petroleum reserves; they dwarf our atomic energy reserves; they dwarf the energy of sunlight itself! Perhaps our children, or their children, will obtain power for their ever expanding needs from the cosmic rays of outer space! Perhaps the man-made satellites of tomorrow will be devoted to the advancement of human progress by bending the energy of cosmic radiation to the will of man.

Which birds fly the fastest?
The hawks, as a group, seem to be the swiftest of all the birds. Of these, the peregrine falcon or the duck hawk is very likely the fastest

of them all. These, in the pursuit of prey, can reach speeds between 170 and 200 miles per hour! There are probably other birds, as yet unclocked, which attain about the same maximum speed. Among the smaller birds, the swifts seem to be the fastest, reaching a top speed of close to 170 miles per hour. Most of the common small birds manage to fly at speeds of 45 or 50 miles per hour. Doves and pigeons do a little better, attaining speeds in the vicinity of 65. Ducks and wild geese are only slightly faster, darting along at 70 miles per hour in level flight. All of these speeds, of course, are maximums; most birds usually loaf along at much slower rates.

While on the subject of birds and their flight: Is it true that some birds can fly under water? As a matter of fact, yes—if we agree that underwater motion and the flapping of wings constitutes flight. Several small sea birds, such as the murres, manage it quite well. A small thrush known as the ouzel makes his living by "flying" around from point to point on the bottom of streams and brooks. An ouzel walks about on the bottom of a brook quite as well as many birds do on land. When he so desires, the ouzel merely flaps his wings and "flies" to a new underwater location.

What is the basis of life?

Life has its basis in *protoplasm*—the stuff of life itself. Protoplasm, in turn, is usually organized into small operational units called cells. Protoplasm is the essential part of each cell which controls all of the processes of life—the absorption of water and chemicals, the manufacture of food, the processes of respiration, growth, reproduction, digestion—in short, all of those functions which constitute "life." It is the only part of a living creature that is really alive.

Biologists have made intensive studies into the physical properties and chemical composition of protoplasm. They know that it is usually a rather thick liquid, somewhat elastic, and resembling egg white in its visible properties. It varies considerably in different organisms and in different parts of the organism. When viewed under a high-power microscope, it appears as a clear liquid (hyaloplasm) in which are suspended minute particles of different size, shape and quantity. These particles are dwarfed by the smallest particle of sand or silt, and even a small bacterium looks large next to one of them. It is believed that this mixture of liquids and particles fills the space between a meshed framework of chainlike protein molecules which are all connected together. Water makes up most of the weight of this living material—often up to 95 per cent of the total. Fats, sugars and mineral salts are also present in varying amounts. The most important components of protoplasm, of course, are proteins, which are composed of carbon, hydrogen, oxygen, nitrogen, sulfur and phosphorus. These compounds are the key substances of life because they form the structural framework of protoplasm and are involved in most of its life activities. The great variety of proteins, their complexity and instability, are undoubtedly important factors in the behavior and existence of the many kinds of living material.

Of course, the really important consideration is the "living" nature of protoplasm. Why is it alive? Just what is "alive" in it? Is it the water? The protein molecules? Perhaps the suspended granules of matter? Unfortunately, when each of these components of protoplasm is analyzed, it appears to be quite lifeless out of the cell. It is evidently their association and interaction which gives life to the whole. The nature of this association, or system of life, is still a mystery. Only one fact seems certain—protoplasm can be produced only by previously existing protoplasm. It does not arise of its own accord from lifeless elements.

The nature of living protoplasm remains the basic problem of biology. Some believe that it differs only slightly from the nature of our inanimate world of chemicals and test tubes. Others believe that living substances differ in some significant and as yet unknown way from the simpler forms of matter. But whatever one's personal attitude in the question, it remains a baffling and absorbing problem —one on which science will continue to exert the force of rigorous experimentation.

Do porcupines throw their quills?

Well, yes and no. No, they can't throw them if throwing implies the voluntary shooting of quills at a target. But you can still manage to get hit with one even though you are standing ten feet from a disturbed porcupine! Like most animals, an angry porcupine will lash its tail and make a furor if it believes that this will improve his situation. In the process, a quill is quite likely to come loose now and then and go flying off into space. The scientific fact that porcupines can't throw quills is small comfort to the unhappy target in such a situation—accidental though it may be. It's safest to allow a reasonable area to the disturbed porcupine, even though he *really* doesn't throw his quills.

Does light have weight?

We have mentioned elsewhere that the true nature of light is one of the most profound and baffling problems facing physicists today. Just what can it be? Does it have weight? Can it be matter in motion? Science came a great deal closer to the solution of this problem as a result of the work of Einstein. He reasoned that light is radiation; radiation is a form of energy; energy has mass; and mass (the substance of the universe) is influenced by the force of gravity. As a result, light passing through the universe should be attracted by the various heavenly bodies just as though it were a miniature planet moving at the speed of light. In other words, if light has mass, it should be bent off course each time it happens to come close to a heavenly body. Einstein proposed a gigantic experiment which would prove the validity of his hypothesis. A solar eclipse was to occur in May, 1919, when the moon would pass between the earth and the sun. This would darken the sky in the daytime and stars close to the sun would be clearly visible. If his theory were correct, these stars would not show up in their normal positions, but would be displaced slightly because of the gravitational pull of the sun on their light as it went past. Photographs of the stars that appear to be in the immediate vicinity of the sun would show a dislocation of their normal position. The Royal Society in London accepted the challenge and sent expeditions to Brazil and New Guinea, where the eclipse would be total. Photographs of the eclipse were taken showing the position of each star close to the sun. Six months later, with the earth at the other end of its orbit, the same stars were photographed as nighttime stars since the sun was then in the opposite part of the sky. When the two photographs were compared, it was found that a displacement had actually taken place—but not exactly as Einstein had predicted. Not one given to discussion, his characteristic reply was, "If they take better pictures next time, the stars will be in their proper places." He was right, of course. After several attempts, astronomers finally performed the experiment with great accuracy in 1952 and the stars showed up "in their proper places." As a consequence of these experiments, we must conclude that light has mass and, therefore, weight!

Do elephants have graveyards?

Probably not; at least no one has ever found one and told about it. But before we close the door completely, perhaps we ought to note the story about Surgeon Commander Levick and his seals. On an Antarctic expedition, near the Drygalski ice barrier, Levick came upon the closest thing to a seal cemetery that one might ever expect to find. A great number of frozen and mummified bodies were spread out in a group, as though they had selected that particular place to die. To Levick, it appeared that the group had slowly accumulated over hundreds of years. There is, at present, no explanation for his find.

And then there's the problem of the missing penguins. Why is it that bodies of dead penguins never seem to turn up? Perhaps the story of Dr. Robert Murphy of the American Museum of Natural History will help to clear up this problem. One day, in the desolate and frozen land of South Georgia, he came upon a pool of snow water. Several penguins, apparently exhausted and weary, stood near its edge. When Dr. Murphy walked to the edge of the pool and looked down into the crystal-clear depths, he saw great numbers of dead penguins. Was this the answer to the riddle?

Do elephants have graveyards? Probably not, but let's not say no —at least not yet.

How do ice floes differ from icebergs?

In the bitter cold of the northern winter, the Arctic Ocean freezes over to a depth of ten feet or more. When warm weather comes, this melting ice cracks up, forming gigantic cakes called *ice floes*. Carried south by the Greenland and Labrador currents, this long procession of snow-covered floes melts in the mid-Atlantic. Except for the snow, these cakes are made entirely of salt-water ice. They rarely exceed ten feet in thickness.

Icebergs are immense, irregular blocks of ice that break off from glaciers and float away in the ocean. Since glacial ice is derived entirely from snow, icebergs contain no salt. Northern Hemisphere icebergs originate in the ocean-terminated glaciers of Greeland. A good-sized iceberg may measure several miles in diameter, and thousands of feet in height. Most of its volume, of course, is under water. It was an iceberg that wrecked the *Titanic* in April, 1912. The collision occurred near Newfoundland, resulting in the loss of over fifteen hundred lives. Nowadays, the International Ice Patrol, operated by the United States Coast Guard, keeps close watch on icebergs. Small ones are blown up by dynamite; larger ones are charted, and their location and path reported to all ships by radio.

Does the color *red* excite bulls?
Apparently not, since bulls are color-blind. Over and over again, experiments have proved that bulls experience a world of black and white, modified only by intermediate shades of gray. Of the animals tested so far, only the monkeys and apes seem capable of color perception. Why then do bullfighters use red capes with which to antagonize the animals? Perhaps, over the years, they have discovered that red excites the clientele.

What would happen if a comet were to hit the earth?
Comets are truly the vagabonds of the universe. They travel in extremely long, narrow curves that bring them well *within* the orbits of the sun's planets, only to whisk them back out into space *beyond* these planets. Some comets seem to belong to the sun, returning again and again at certain predictable times. Others are true vagrants of the heavens, rushing in to pay us a single visit and then speeding out into the depths of space, probably never to return. But what is the nature of comets? What are they made of? The head, or nucleus, of a comet must be practically empty space! It is believed that it consists of a swarm of relatively small solid particles. These particles are so separated that stars are practically undimmed when a comet's head is in position to shield them from our view. The entire mass of a comet must be quite small. When a comet approaches a planet, its path is altered considerably by the planet's attraction. The planet, however, shows no appreciable change in direction as a result of the comet's presence. If a comet were to strike the earth— the chances against this happening are about fifteen million to one —we would probably get nothing more than a dazzling shower of shooting stars.

Perhaps one of the most amazing characteristics of a comet is the odd behavior of its tail. On approaching the sun, the tail follows along behind, much as we'd expect it to. But when the comet retreats to outer space, the tail goes *in front!* Scientists believe that the tail consists of a stream of gases released from the solid material by the heat of the sun. They also believe that the gases are pushed away from the sun by the pressure of sunlight! For this reason, the comet's tail is always directed away from the sun.

By far the best-known comet is the one predicted to visit the earth "about the year 1758," by Edmund Halley. He noticed the similarity of the orbits of three earlier comets, and suspected that they were really three appearances of the same comet. Just as Halley predicted, the comet appeared on Christmas Day of 1758. It has subsequently returned in 1835 and in 1910. Twenty-eight visits of this comet have been identified from records dating back to 240 B.C. It is expected to visit us next in the year 1986. At the present time, it is invisible at the far end of its orbit beyond the planet Neptune.

Can cats see in the dark?
No, but they do have extraordinarily good vision even in the dimmest light. They can see much better in such light than we can. Why, then, aren't they blinded by the brilliance of full sunlight? Unlike humans, with our round pupils, cats have slit pupils. They can draw these slits tightly together, in sunlight, so that only the tiniest amount of light enters. When the light subsides, the slits open up to allow a great deal of light to enter.

Why does the setting sun look oval in shape?
Have you ever noticed the apparent break in an oar that's partly submerged? This effect is an optical illusion caused by the bending, or *refraction*, of light rays when they pass at an angle from one medium (water) to another (air). The sun looks noticeably flattened when near the horizon for just this reason. Light coming from the sun must pass from airless space into our gaseous atmosphere. In doing so, sunlight is bent down toward the earth in a gentle curve. This bending depends upon the angle of the sun in the sky; the lower the position of the sun, the greater the bending of its rays. When the sun is directly overhead, its light is not bent at all but comes through the atmosphere in a straight line.

Now let's visualize the setting sun, just before it slips below the horizon. A ray of light coming from the top of the disk is bent down toward the earth a certain amount because of refraction. This produces an optical illusion. It makes us think that the top of the sun is *somewhat above its actual position*. This is important so let's try an experiment to illustrate the point. Place a small mirror in a horizontal position with its reflecting surface down. Hold it at arm's length, slightly above eye level so that you can see the image of an object on the floor. If you didn't know better, you would think that the object was up on the ceiling rather than on the floor. This is caused by the bending—reflection is just another kind of bending —of the ray of light coming from the object. Your eye sees the ray

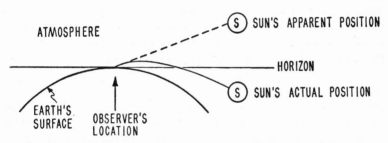

FIG. 24. REFRACTION OF SUNLIGHT BY THE EARTH'S ATMOSPHERE
The earth's atmosphere becomes less dense as the altitude increases. This causes sunlight to be bent, or refracted, in a curve (exaggerated) as shown above. This effect increases the length of our day by several minutes.

coming from the direction of the mirror and assumes that it started out from that direction. We know that the object is on the floor but it *seems* to be on the ceiling. The eye incorrectly assumes that light always travels in straight lines. It notes the direction from which the ray is coming and assumes that it started out from its source in a straight line. This is why your image in a normal mirror seems to be behind the mirror rather than in front of it, where you really are.

Now to get back to the flattening of the setting sun. We have already pointed out that the top of the sun seems to be somewhat higher in the sky than its true position. This is because your eye sees the ray of light *after* it is bent down toward the earth by refraction. You assume that the ray came to you in a straight line, whereas it had, in fact, been bent down by our atmosphere. The same thing happens to a ray coming from the bottom of the sun, but even to a greater extent. Such a ray is bent a bit more than its upper neighbor because it is lower in the sky. This means that the bottom of the sun is "raised" a greater percentage than the top. This apparent compression makes the sun look flat or oval-shaped. The magnitude of the sun's apparent height increase is quite appreciable. The sun on the horizon is raised above its true position by an amount slightly greater than the apparent diameter of the sun or full moon.

Do butterflies migrate?

For many years, scientists and nature lovers have wondered about the behavior of a few of the fragile-winged insects that appear to travel south in winter just as migratory birds do. Typical of these insects are the showy, black and orange monarch butterflies which live as far north as Canada. In fall they gather in great hordes and fly south, as if to summer in Texas or Florida. Do these small insects have the brains and endurance to select such a course of action in order to avoid the northern winter? In order to find out, Dr. Frederick Urquhart, director of Toronto's Museum of Zoology, embarked on a nineteen-year investigation of the monarch's habits. He began in 1938 by trying to attach labels to monarchs. He hoped by this process to find out how far they flew. His first attempts met with failure, however, as the paper labels tended to make the butterflies aerodynamically unstable. During the war, Dr. Urquhart became familiar with Royal Canadian Air Force planes and his newfound knowledge was put to use in creating a better tagging procedure. He found that small paper labels attached to the leading edges of a monarch's wings worked out very well. Holes were cut in the butterfly's wings close to the front, so that the label could be bent over and glued to itself rather than to the virtually glueproof wing. In 1950 and 1951, Dr. Urquhart tagged three thousand butterflies, set them free, and waited for replies to the address tags. Unfortunately, his efforts again met with failure. The tags were coming loose in wet weather. When the next migratory season rolled around, Dr. Urquhart was ready with waterproof labels. In 1955, tagging was begun on a large scale; ten thousand monarchs were tagged by hundreds of aides. This time only partial success was realized. Tagged monarchs were found only as far south as Virginia. This turned out to be the work of predatory birds that seemed to enjoy eating tagged monarchs much more than their untagged relatives.

The piece of white paper seemed to destroy the insect's normal camouflage. Undaunted, the doctor and his aides tagged thousands of additional monarchs in 1956. As a result, insects tagged in Ontario were found all the way down to Texas and the Gulf Coast. At last, success. It seems that several generations of the insects live and die each summer in northern climates. The generation that is alive when cold weather arrives migrates south to spend the winter. Since monarchs don't reproduce in the south, the same insects fly back north in the spring.

Within the past few decades, monarchs have been turning up in England in ever increasing numbers. How do they get there? The results of Dr. Urquhart's untiring efforts have given us a pretty sure clue—they fly!

Why do roots grow down into the ground instead of up?

We know that stems ordinarily grow upward toward the sky while roots tend to grow into the earth, or at least laterally. While this state of affairs is quite "normal," we may still wonder about the factors that influence these directions of growth. After all, the plant can't "know" which way to grow—and even if it did, it couldn't respond to that knowledge with action. If a young seedling is transplanted with its root and stem in random positions, the root will invariably turn its tip downward while the stem begins to grow upward. More than a century ago, the English horticulturist Knight suggested that this change in direction of growth was due to the force of gravity. He reasoned that if this were so, it ought to be possible to substitute a greater force for gravity and thereby cause the plant to grow in other than normal directions. In order to accomplish this purpose, he fastened young plants in various positions to the rim of a wheel which he could spin rapidly in a horizontal plane. This subjected the plants to a centrifugal force much greater than the pull of gravity. This force, of course, was outward tending to pull the plants from the wheel. After a reasonable length of time, Knight examined the plants and found that the plants had done just as he expected; the roots had grown outward, in the direction of the centrifugal force, and the stems had grown inward toward the hub, in opposition to that force.

In a similar experiment, the wheel is slowly rotated in a vertical plane. This exposes all sides of the seedlings to the pull of gravity

221

without substituting a large centrifugal force in its place. Under these conditions, the pull of gravity on the plant has no net effect, since it tends to pull the plant first one way and then the other. As a result, the root and stem continue to grow in the direction in which they happened to start; they show complete indifference to their normal directions of growth.

The sensitivity of plants to gravity is called geotropism. If the tip of a root, for example, finds itself pointed in the wrong direction, it *grows* around into the proper direction. This is accomplished by the selective distribution of growth-stimulating substances called *auxins*. If a root tip is in a horizontal position, more auxin is sent to the top than to the bottom of the tip. This causes the top to grow faster, resulting in the "bending" of the root downward. In this case, a preponderance of auxin circulates to the top of the root, *against* the pull of gravity. When stems bend upward, just the-opposite happens. Auxins congregate on the *lower* portions of the stem *toward* the pull of gravity. Why this differentiation exists, and why auxins circulate in such a manner at all, are questions still facing the science of botany.

Which animal lives the longest?

There are about as many answers to this question as there are authorities, but perhaps a concensus will be of interest. In general, large animals live longer than small ones; vegetarians and animals that eat mixed foods live longer than meat eaters; animals that mature slowly, also live longer; and cold-blooded animals live longer than hot-blooded ones. The most reliable information available on the longevity of animals comes from the zoos. From this source we find that, among the primates, the life span seems to range from about ten to forty years. The common rhesus monkey lives to the age of thirty while the chimpanzee can reach the ripe old age of forty. The American beaver is the patriarch of the rodent family, often living to the age of nineteen years. His cousin, the mouse? Perhaps four or five years. Among the cats, the lion is still the monarch, living to be thirty years of age.

Do elephants live for hundreds of years? By no means. Most elephants are quite rickety by the time they reach fifty or sixty. How about fish? Carp live to be about sixty or seventy while such smaller ones as perch, bass and trout rarely exceed fifteen years. The range of life among the insects is an oddity in itself. While the May fly and thousands of other varieties measure their span of life in hours, termites and bees may live between fifteen and twenty-five years! Some of the birds will also amaze you in their ability to withstand the wear and tear of the years. Parrots, owls, ostriches and eagles seem to live the longest—sometimes as long as we humans. Bats do very well for their size, living to be ten to fifteen years of age. Cats and dogs? About twenty and fifteen years, respectively.

Which animal lives the longest? No one seems to know for sure. Some think the giant tortoise may hold the prize, living to the age of two or three hundred years. But little authenticated data is available to prove this conclusively. Even if it were true, the tortoise might have competition. Very little is known concerning the age of such animals as the whales, porpoises, kangaroos, anteaters and duckbills. Perhaps one of these may earn the title. There is only one way to find out, and that takes patience—and time.

What were the first tools made of?

Aside from the use of sticks and clubs, the first tools appear to have been made of bone. It was available in large quantities; it was reason-ably strong; it was harder than wood; and it was easy to work with. But another and more durable material soon took its place—stone. And early man's experience soon told him that flint was the best kind for the purpose. Flint knives are among the earliest known tools. They enabled man to make better use of all of the previously known tools. He could sharpen spears for hunting, make better hatchets for chopping, and produce implements of wood and bone for an ever increasing number of activities.

With the ability to develop a keen edge, a sort of industrial revolu-tion may have taken place. For the first time, man had a weapon that made him the master of his environment. This must have provided him with time for relaxation for at about this time the first prehistoric art began to appear. Skillful drawings of animals were cut on fragments of wood, bone and stone, and on the walls of caves. Man had not only developed a keen edge, but also a mind that was intelligent enough to create and appreciate a primitive form of art.

Does evolution imply that we are descended from apes?
The subject of man's relationship to the animals has evoked considerable contention over the years. This has occurred in spite of the practically universal acceptance of evolution by naturalists and biologists. Evolution, as we have discussed elsewhere in this book, contends that present forms of animal life have come into existence as a result of changes in pre-existing types. Scientists believe that the first creatures on earth were not the foxes, dogs and horses of today, but simpler animals, similar to the single-celled Foraminifera whose skeletons make up our chalk cliffs and marble. From several such forms of life—or perhaps even from one—have evolved all of the forms of animal life on earth. With their amazing ability to adapt to changing circumstances, living things have variously altered their sizes, shapes, coloration and habits; this has taken place over tremendous periods of time—slowly, gradually, effortlessly. The forms in existence today have survived by changing to meet the tests of an ever changing environment.

Is this theory hostile to religion? Scientists say that it isn't. Most writings on the subject have emphasized this point. Of course, it is hostile to a too literal interpretation of the Book of Genesis. Genesis speaks in terms of days, and evolution interprets them as extremely long days indeed. Genesis speaks of the divine origin of life and things, and about this, the fundamental substance of religion, science never comments, one way or the other.

Now to get back to our alleged ape ancestry. Naturalists seem to be in agreement that we are *not* descended from apes, monkeys or any other existing primate. Evolution indicates that back in the dim geological past, man and ape branched off from a common ancestral form of life. Ultimately, with the passage of age upon age, the two forms of life came to exist in their present condition. Are we, then, related to the apes? According to evolution, yes. We are related to the apes, the horses, the sheep and all other animals. True, the relationship may be remote, but according to the theory of evolution, it exists nevertheless.

Is blood alive?
Every portion of our body is irrigated by a complex network of human plumbing—the blood stream. It is the route by which the human heart pumps about a gallon of blood throughout the body.

It provides fuel and oxygen to living tissue, and removes the waste products that would detract from their efficiency. And in addition to this, the flow of blood is itself a living stream. Living cells, both red and white, flow in a liquid stream throughout the body. The red cells, known as *red corpuscles*, carry oxygen to the body and eliminate some waste products. The blood of a normal person contains billions of them! If their number is reduced, *anemia* results, and the body suffers from a deficiency of oxygen and a burden of body poisons.

When the body is attacked by a foreign microbe, it is the *white corpuscles* that come to our rescue. They engulf and destroy the bacterial parasites that try to take life easy at our expense. If the red corpuscles can be thought of as *workers*, then the white cells are the *soldiers* of the body.

Both the red and white corpuscles circulate throughout the body in a river of chemicals known as *plasma*. Besides providing transportation for the working and fighting cells, plasma carries dissolved nutrients to the hungry muscles of the body. It also provides the mechanism by which blood coagulates, or plugs a wound in the body. Were it not for this unique ability of plasma, all of us would suffer from hemophilia, a disease which prevents the blood from clotting. But as in all of the delicately balanced processes of life, the coagulation of blood can also be a liability. Ordinary heart attacks, such as coronary thrombosis, result from a blood clot which obstructs the flow of blood, causing paralysis or even death. When we consider the complexity of the human body, we can only be amazed at the nice sense of balance it displays between death-dealing extremes; we can only wonder at its ability to make so few mistakes.

How well can birds see?

Excellence of vision means more to birds than any other sense. The expression "eagle-eyed" pays due tribute to their uncanny ability to locate small creatures that move around far, far below. Have you ever noticed the apparent ease and great precision with which the various birds negotiate a landing on a narrow ledge, or even the wires of a telephone pole? Have you ever watched a bird swoop past the branch of a shrub, deftly picking off an insect on the way? These and other everyday activities of birds indicate extremely keen vision. Scientists tell us that the pigeon has such fine visual ability that it can distinguish a caterpillar six hundred feet away! Another case on record illustrates the keen sight of the eagle. One day the ornithologist E. H. Eaton was standing on the shore of a lake watching an eagle soar high in the sky. Suddenly the eagle made an unswerving, diagonal dive toward the shore of the lake, where it seized a fish that it had detected. The point at which the catch was made was so distant that binoculars were needed to assure the scientist that the fish had been caught. When the distance was measured off, it was found that the eagle had begun his dive three miles away from the point of seizure!

What are asteroids?

The nine known planets that revolve around the sun are usually divided into two groups: the minor planets, consisting of Mercury, Venus, earth and Mars; and the major planets, consisting of Jupiter, Saturn, Uranus, Neptune and Pluto. This distinction is made, not on the basis of size, but on the basis of their distance from the sun. The minor planets are closest to the sun; then there is a large gap, and then the major planets begin. Mercury is the closest to the sun, followed by the others in the order given above. Although there is a wide interval between Mars and Jupiter, the space is far from empty. On the contrary, it is filled with thousands of very small planets called *asteroids*. Ceres, the largest of the group, has a diameter of only 480 miles. The smaller ones are no larger than big stones.

The asteroids revolve around the sun just as do their larger planetary neighbors. Some of their orbits, however, are extremely erratic. A few of them wander in an elliptical path so elongated that it brings them inside the orbit of Mars and outside the orbit of Jupiter. One of them has even been known to approach to within three million miles of the earth. As great as it sounds, this distance is merely an astronomical hop, skip and jump. Fortunately the great majority of these celestial wanderers keep to their own portion of the solar system—the region between Mars, the last of the minor planets, and Jupiter, the first of the major ones.

What is gossamer?

Most of us associate the spider's silk-spinning activities with web construction, but it turns out that spider silk is used for a variety of purposes. The spinnerets located on the spider's abdomen provide fine silk for the fabrication of egg sacs, coarser silk of many strands for the binding of prey, sticky silk for trapping purposes, and a filament, called the *dragline,* which is used for communication and transportation. The dragline is usually spun continuously as the spider moves around. It's the dragline that enables a spider to raise and lower himself when we find him dangling in space, and this same filament provides aerial locomotion when a strand is caught by a fast-moving breeze. In the course of time, spiders spin a great amount of dragline silk for which they have no further use. Gossamer is merely a film of draglines, floating delicately through the air under the influence of a gentle breeze.

Before we get off the subject of spiders, someone will want to know about cobwebs. These are merely draglines that have found a more or less permanent home.

Do we use numbers backward?

The number 746586970964 is a pretty difficult number to read when we proceed in the normal direction, from left to right. Now try it from right to left. It gets easier, doesn't it? Four units, 6 tens, 9 hundreds, and so forth. Now let's divide the number with commas, again proceeding backward, from right to left: 746,586,970,964. After translating the number into English, we can now read it as seven hundred and forty-six billion, five hundred and eighty-six million, nine hundred and seventy thousand, nine hundred and sixty-four. We read numbers backward because we derive our numerical system from the Arabians; and the Arabians do all of their reading and writing from right to left.

If you're still a bit skeptical, try adding the following numbers:

$$891424$$
$$\underline{93628}$$

Once again, we find the sum like the Arabians do, from right to left. Now try subtracting the numbers given above. Or try multiplying them. In every case we fall back on the Arabian method which is just the reverse of the way we read words.

Who were the first to use a fishline?

In a book of this kind, the author would hardly be human if he could resist the temptation to include at least one "trick" question.

But this one is so unusual that perhaps the reader will find it possible to overlook the first offense.

Perhaps the most ugly and uninviting creatures on earth are our ordinary earthworms and angleworms. Even the most ardent fisherman will agree that they are far from attractive. The worms of the sea are another matter, however. Many of them, unlike our common earthworm, look like beautiful, flower-like growths rising from the sandy bottom. They often display a rainbow of colors, gleaming with brilliant emerald, gold, blue, crimson and purple. Minute scales on the worm's body act as prisms, breaking up the sun's rays into beautiful, iridescent colors.

Some sea worms, while not as beautiful as their colorful relatives, are nevertheless strange and interesting in their habits. One of these is the bootlace, or fishline worm. They are quite common and resemble small pieces of orange or yellow liver lying under rocks and stones. If you place one of these creatures in a pail of water, he will amaze you by immediately uncoiling into a whitish, slender worm of apparently limitless length. They have been found to measure eighty or ninety feet in length! As if this weren't spectacular enough, these worms are equipped with sucking, "fishhook" mouths that attach themselves to small fish with relentless tenacity. Once "hooked," the fish is never released, but is "played" by the resistance of the worm's body through the water. After the fish becomes exhausted, it is devoured by its odd captor. I think you will agree that this is one of the strangest fish stories that you have ever heard.

How does the telephone send sound over wires?
The invention of the first practical telephone is generally credited to Alexander Graham Bell in the year 1876. To understand how the telephone works, it will help to recall that a sound wave consists of a series of disturbances set up in air by a vibrating body. When a violin string vibrates, it pushes alternately against the air on one side and the other. The resulting disturbances move away from the string in much the same manner as water waves produced by a stone thrown in the water. The string produces a series of disturbances that we call sound.

Our modern telephone handset consists of two essential parts: a receiver and a transmitter. The transmitter changes sound waves into variations in electrical current and the receiver reverses the process at the other end of the telephone line. The transmitter consists of a small container filled with lightly packed carbon particles. A diaphragm on one side of the container moves back and forth in accordance with the vibrations of sound waves that strike it. Wires are connected to the terminals of the container so that an electric current passes through the carbon particles. When a sound wave strikes the diaphragm, the disturbance causes a varying pressure to be applied to the carbon particles. When they are pressed closer together, they touch in more points and the electric current goes up. When they are allowed to spread apart, they touch in fewer places and the current goes down. This variation in electric current through the carbon particles corresponds closely with the vibrations of the sound wave.

Once the sound vibrations have been changed into a varying electric current, it is possible to send the current over a telephone line to the receiver at the other end. The receiver consists of fine, insulated wire wound around the poles of a small magnet. A thin iron diaphragm is mounted so that it is near the poles of the magnet. When the current variations reach the coils, they vary the strength of the magnet. This causes the diaphragm to be attracted to a greater or lesser extent depending upon the strength of the magnet. The iron diaphragm must vibrate, therefore, in accordance with the electrical variations passing through the coil. This causes the diaphragm to reproduce the sounds that caused the current variations at the transmitter.

Is there an easy way to pull a car out of the mud?
The study of forces brings to light many unexpected facts. Take the case of a marble rolling down an incline, for example. We know that gravity pulls the marble straight down and not to the side. Why, then, does the marble move at all? Scientists answer this question by showing that any force may be broken down, or resolved, into a number of individual forces. These forces, when added up properly, equal the original force. The inclined plane, mentioned above, resolves the force of gravity into two forces at right angles: one acting in a direction down the slope, and the other acting at right angles to the slope. Similarly, none of us pushes a lawn mower in a truly horizontal direction. This would require our bending down until the handle was as low as the wheels. Instead, we push the mower at

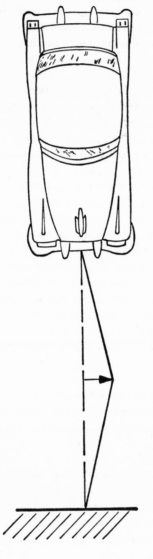

FIG. 25. AMPLIFICATION
OF FORCE

One end of the rope is attached to the automobile. The other end is drawn tight (dotted line) and attached to a distant stationary object. When the center of the rope is pushed to the side, this force is amplified and pulls the car forward.

an angle to the ground. The pushing force is resolved into two separate components: one in a horizontal direction along the ground, and the other directly into the ground. You can prove this to yourself by trying to push the mower with the handle almost vertical. Under these conditions, the mower doesn't move forward very fast and most of your effort is expended in making indentations in the ground.

In many situations of this sort, the component forces produced are smaller than the original force itself. It's possible, however, for a force to produce components in useful directions which are considerably greater than the original force. One such method is useful for pulling stuck cars out of the mud or sand. Attach one end of a long rope to the front bumper of the car and pull the other end tight. Anchor this end to a distant tree or other solid object. Then take the rope by the middle and walk a few feet to the side. The force you apply is multiplied and moves the car toward the tree. The brakes are then set by an assistant and the rope tightened for the next pull. The process is repeated until the car is free of its predicament. The force produced by this method is many times that applied at the center of the rope. Since we can't ever get something for nothing, we find that the forward motion of the car is smaller than the sidewise motion of the center of the rope. It's smaller in the same proportion as the force on the car is larger. The greater the force amplification obtained, the smaller will be the forward motion of the car.

This principle is used in reverse by tightrope walkers. The rope in this case is intentionally given some slack so that it will not break. If the rope or wire had no slack at all, great forces would be developed in the rope by the weight of the performer's body.

What causes the "man in the moon"?
From very early times, the dark markings on the full moon have been likened to a man's face. An early legend tells us that the man was put there to gather sticks on Sunday. Depending upon your own particular brand of imagination, you might see the "lady in the moon," the "girl reading a book," the "crab" or the "donkey." Some have even fancied a sitting "French poodle" that is trimmed in the conventional manner, even to a distinctive "pompom" at the end of his tail! In the early days, it was supposed that the moon's markings

were merely reflections of the oceans and continents of the earth. In 1610, however, Galileo looked up at the moon through his new telescope and saw for the first time that the markings were configurations on the surface of the moon itself. He concluded erroneously that the dark markings were seas, and they have come to be known by the Latin name *maria*. Each "sea" was given a name that was thought to be in keeping with its appearance, such as *Mare Imbrium*, "the Sea of Showers," and *Mare Serenitatis*, "the Sea of Serenity." Today it is known that there is no water on the moon. The *maria* may have been seas at one time, but they are now nothing but large, desolate plains.

In addition to the fancy of the observer, there is another logical reason to explain the variation in the figures seen on the face of the moon. The configurations of the full moon don't maintain the same position at all times and from all points on the earth. They can appear in vertical, horizontal, reversed, and in all intermediate positions depending upon the latitude of the observer and the angle of the moon. The "man in the moon" looks upside down to people in Argentina. This is because Argentina is more than twenty-nine degrees below the Equator, the point at which the "man in the moon" begins to turn over on his head. It's only natural to find, therefore, that people in different parts of the world discover different figures in the markings. To any observer, the "top" of the moon is that portion that happens to be farthest from the horizon. The part of the moon that forms the "top" will be different as seen from various parts of the earth. You can illustrate this point by studying the moon from a lying position. If you let your imagination run loose, there is almost no limit to what you can find!

How dangerous is the radioactive fall-out from nuclear explosions? This problem is extremely complex, requiring a knowledge of medicine and genetics as well as physics. But scientists are beginning to get to work on it. Of all of the products of nuclear explosions, strontium-90 is probably the most feared. It's similar in chemical properties to calcium and is absorbed along with calcium by the human system. Once in the body, it is deposited in the bones where its persistent radioactivity may eventually cause cancer.

In order to find out how bad the situation really is, the United States Atomic Energy Commission financed a study by Drs. J. L.

Kulp, Walter R. Eckelmann, and Arthur R. Schulert of Columbia's Lamont Geological Observatory. In February, 1957, this team made its report. From five hundred samples of fresh human bone from many parts of the globe, the team concluded that strontium-90 can be found in all human beings, "regardless of age or geographic location." The amount is not too large. Averaging the results of all of their analyses, they concluded that the human race now has 0.12 micromicrocuries of strontium-90 for each gram of body calcium. The *micromicrocurie* is a measure of radioactivity equal to 0.037 atomic disintegrations per second. According to their results, the average human being has about one-thousandth of the presently accepted maximum permissible concentration.

Despite this reassuring note, the research team noted a large variation between individuals. Young children averaged three to four times as much strontium-90 as adults. Even among the adults there was great variation. One adult individual had seventy-five times as much of the radioactive isotope as his fair share.

Strontium-90 is not normally found in nature. It's a product of the giant thermonuclear explosions of the United States, Russia and England. After an explosion it rises high into the stratosphere and from there it falls slowly until it reaches the lower atmosphere. It is then quickly washed to earth by rain. All of this takes time; strontium-90 makes use of this time by spreading over the face of the earth. When it gets into the soil, it is taken up by plants as if it were calcium. Some plants seem to like the radioactive isotope more than others, as is shown by the tests of Japanese scientists. The Japanese sampled a long list of substances, from spinach to deer horns, and found wide variations in the strontium-90 content. Tuna caught near Bikini atoll had the highest radioactivity: 53.5 micromicrocuries. Tea contained 30 while spinach contained only 3.8. Rice, an all-important food in Japan, contained the relatively high count of 10.4. The scientists also analyzed the ashes of twenty persons taken from burial urns. Their strontium-90 count varied from 0.06 units to 4.1.

Animals that eat plants containing strontium-90 seem to reject it in favor of calcium. This means that milk contains less of the isotope in proportion to calcium than the plants that cows eat. Humans that get most of their calcium from milk are therefore better off than people who get it directly from vegetables.

Still to be investigated are the many other fission products of thermonuclear explosions. What effect does radioactivity have on genetics? Why does one human absorb so much more of the isotope than another? And how about the permissible level itself? It was derived from a small amount of experience with radium and cancer. At that time there was no strontium-90 in the world. When more work is done, the permissible level may have to be reduced sharply. In the words of Dr. Masao Tsuzuki, director of Tokyo's Red Cross Central Hospital: "Speaking as a scientist, I can make no evaluation of the strontium-90 danger. Too much work is still to be done. We do not know how much gets down to earth or how long that takes. We do not know how much then enters the human body, or at what rate, or what the mechanism of transfer from food to animals and humans is. I do not believe that strontium-90 will be permanently harmful at its present level, but if experimental explosions continue at the present rate, there will come a time when the human body will be seriously harmed. It will then be too late to do anything about it."

What is a lever?
A lever is a rigid bar that can turn about some point called the fulcrum. It has been in use for countless centuries. It was probably first used to dig up edible roots or to move heavy logs out of the way. It has since been developed into oars, paddles, shovels, crowbars, scissors, wheelbarrows, nutcrackers, beam balances and a wonderful variety of other tools and appliances. Even the human forearm is a lever! In more specific terms, a lever is a bar upon which a force may be applied to overcome a resistance or load. The bar is pivoted at the fulcrum around which it moves when the force is applied. A seesaw is one of the more pleasurable forms of the lever. If the children are of equal weight, they sit at equal distances from the fulcrum. If one is heavier than the other, the heavier child must sit closer to the fulcrum. This illustrates the principle of the lever. The weight of one child times his distance to the fulcrum must equal the weight of the other times his distance to the fulcrum. The same law applies to the simple beam balance. If we suspend a yardstick by the 18-inch mark (the fulcrum) we have constructed a balance. Suppose we want to use this device to determine an unknown weight. All we need do is suspend a known weight, such as a pound of butter, on one side of the fulcrum, and the unknown weight on the other. We can balance the beam by moving the two weights around. Let's

assume that the beam is balanced when the 1-pound weight is at the 3-inch mark, and the unknown weight is at the 6-inch mark. The principle stated above tells us that the known weight (1 pound) times its distance (3 inches) must equal the unknown weight (?) times its distance (6 inches). Solving this problem tells us that the unknown weight is equal to ½ pound.

A crowbar is able to exert such great force because of lever action. The device is constructed so that the fulcrum is only a short distance from the weight to be moved, perhaps only an inch or two. The

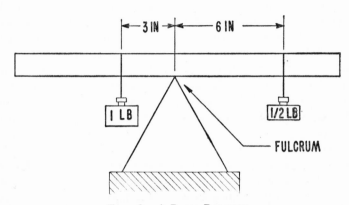

FIG. 26. A BEAM BALANCE

This device will be balanced whein the weight on one side times its distance to the center equals the weight on the other side times its distance to the center.

applied force, on the other hand, is usually a yard or so away from the fulcrum. This enables a small force applied at one end to balance and move a much larger resisting load at the other. A heavy load can be lifted in a wheelbarrow for the same reason. In this case, most of the load is borne by the wheel.

Can any plant subsist on solid rock?
The *fungi* are a family of rather simple plants that are completely lacking in chlorophyll, and are therefore unable to manufacture their own food. They obtain organic food material from one of two sources: from living plants and animals, if they are *parasites;* or from the partially decomposed remnants of dead plants and animals, if they are *saprophytes.*

Most of the parasites disturb or interfere with the normal life processes of the host, resulting in a diseased condition. Many of the plant diseases, such as the mildews, smuts and rusts, are a result of this type of fungal growth. In those rare instances where fungi affect animals, the disease is usually in the form of a skin infection, such as in "ringworm" or "athlete's foot." Treatment of fungal plant diseases is usually preventive rather than curative. It consists largely of spraying with chemicals that inhibit the normal growth of fungal tissue.

Fungi of the second, or saprophytic, group consist mainly of the molds, yeasts, bracket fungi and mushrooms. The molds cause fermentation in food products and other organic materials; the yeasts bring about fermentation in sugar solutions; the bracket fungi aid in the decay of dead trees and plants; and the mushrooms and toadstools provide us with both poisonous and edible forms. The yeasts are probably of greatest economic value to man. Through the process of fermentation, they are able to break up sugar into alcohol and carbon dioxide gas. If certain strains of yeast are placed in grain extracts, or in fruit juices, the normal sugar content is changed into alcohol, and beer or wine is formed. When yeast is mixed with bread dough, the carbon dioxide gas that is generated causes the dough to "raise" since it finds it impossible to escape from the rather thick mass. What happens to the alcohol that is formed in the dough at the same time? The heat of subsequent baking causes all of it to evaporate.

While of secondary importance to the yeasts, their fungal cousins, the molds, are also useful to man. Certain molds are used in the manufacture of the tastier or so-called "fancy" cheeses. In some cases they play a more important role in cheese-making than bacteria, the creature that is usually given most of the credit.

The oddest fungi of them all, however, are fungal plants known as the *lichens*. They are the plants that defy nature, growing on rocks, tombstones and other similarly uninhabitable places. Lichens of the Arctic constitute the normal food for reindeer and caribou, and tremendous herds eat the lichens right down to the stones on which they grow. Once the lichens in a particular area are eaten, it takes about eight years for a new crop to grow, a fact that helps to account for the unpredictable migratory behavior of these animals of the north. But the odd part about the lichen is its dual nature; it is really not one plant, but a combination of two entirely different ones. The leathery body is made up of the tissues of a *fungus,* and imbedded in this material are many microscopic green plants called *algae.* While the fungus cannot manufacture its own food, algae can, due to the presence within their tissue of chlorophyll, nature's own food factory. The fungus acts as a parasite, deriving its carbohydrate food from the algae. But this parasitism doesn't seem to adversely affect the algae; it too seems to benefit from the association. In exchange for its food, the fungus provides moisture and protection to the algae. As with many other forms of life, this association is mutually advantageous: you might say that the fungus pays his food bill by providing moisture and protection to the algae. An association of this kind is called *symbiosis.*

In addition to their remarkable mode of life, the lichens play an important role in nature. They are able to grow on bare windswept ledges, or other seemingly impossible locations where no other plant could survive. As a result, they usually constitute the first vegetation in regions previously bare of plants. Growing in such places, they accumulate particles of soil to which they contribute humus through the decay of their own plant tissue. In this way, new soil is formed which is eventually capable of supporting other, less vigorous forms of plant life. To illustrate their capacity for life under severe conditions, there is the experience of Professor L. M. Gould, on his journey southward from the base of the Byrd Antarctic Expedition. On this occasion, he found lichens growing on the bare ledges of Mount Nansen, within five degrees of the South Pole. Lichens, that strange combination of fungus and algae, are truly the most rugged members of the plant kingdom.

How can a magnet produce electricity?

Michael Faraday was born in 1791 of a poor English family. Although he had little formal education, he obtained a position as laboratory assistant (mostly bottle washing) at the Royal Institution in London. Because of his brilliant discoveries and inventions, he eventually rose to become head of this famous institution. Perhaps his most significant achievement was the discovery that a magnet can be used to produce electricity. This discovery laid the foundation for the electrical industry as we know it today.

Suppose we make a coil of wire and connect the ends to a meter that is capable of measuring a current of electricity. The meter would normally read zero since there is no current flowing through the coil of wire. Now let's plunge a magnet into the coil and see what happens. The meter pointer suddenly jumps up to a high reading and then back to zero as the magnet comes to rest within the coil. When we withdraw the magnet from the coil, the pointer moves in the opposite direction for an instant and then returns to zero. The direction in which the pointer moves depends upon which way the current is flowing through the coil. When the magnet is thrust into the coil, the current goes in one direction through the wire; when the magnet is removed, the current flows in the opposite direction. It makes no difference, by the way, whether the magnet moves or whether the coil moves; it's only necessary that there be relative motion between magnet and coil.

Physicists explain this phenomenon by stating that the magnet is surrounded by *magnetic lines of force*. These are invisible lines running from one pole of the magnet to the other. When magnet and coil move with respect to one another, the lines of force are "cut" by the coil of wire. This induces a current of electricity to flow in the coil. An induced current of this kind flows only for the period of time during which the magnetic lines of force are being cut. As soon as the relative motion stops, the current stops.

An electrical transformer is merely a combination of two coils of wire wound very close to each other. They are not connected to each other electrically. When a current of electricity is passed through one of these coils, it produces lines of force in its own right. These lines of force are exactly the same as those produced by a magnet. If the current of electricity is made to vary continually, the lines of force vary in the same manner. They build up and decrease in response to the changing current that produced them. Since these lines of force are in motion, they cut across the second coil of the transformer. This induces a current in the secondary coil just as though a magnet had passed through its center. The current so produced will vary in accordance with the current in the first coil. As you can see, electrical energy has been transferred from one coil to another without the aid of an electrical connection between the two.

In practice, transformers are made of coils that are wound on cores of iron. This concentrates the lines of force so that practically all are useful in inducing current in the second coil. Transformers can increase or decrease the voltage of an alternating current. If a *voltage step-up* transformer is desired, more turns are used on the secondary coil than on the primary. If a *voltage step-down* transformer is needed, less turns are used on the secondary.

Why do violinists bow the strings near one end?

A violin string produces a sound because it vibrates in response to friction with the bow. There are, however, many ways in which a string may vibrate. It may vibrate as a whole, for example, in which all parts of the string move except the very ends. This type of vibration will produce a sound consisting of one frequency, the fundamental tone. A string may also vibrate in two equal parts, in which case there is a point at the center of the string which does not move. This type of vibration is analogous to the vibration of two half-size strings connected end to end. This produces an overtone which is twice as high in frequency as the fundamental. Similarly, if the string is caused to vibrate in three or four equal parts, other overtones will be produced. In each case the sound is different. It depends upon the type of vibration induced in the string by the bow. By bowing the string about one-seventh of the way from one end, a violinist causes the string to vibrate in parts at the same time that it vibrates as a whole. This enriches the quality of the tone by the addition of desirable overtones. In much the same way, the hammers of the piano are located in such a way that they will set up overtones.

How does planting legumes increase soil fertility?

Although our atmosphere consists mostly of nitrogen, this nitrogen is in the natural gaseous form which plants can't use. When combined with oxygen and other elements, however, nitrogen is a valuable and important plant food. *Nitrates,* as such compounds are called, must be present in the soil for plants to use. They can be put there in the form of fertilizers, or by the growing of leguminous plants. These are members of the pea family, a large group of food and forage plants that produce fruit in the form of a *legume* or true pod which opens along two seams.

All legumes have (in varying degree) the peculiar ability to transform or "fix" atmospheric nitrogen into nitrogenous compounds which plants can use. This is accomplished through a remarkable partnership between these plants and certain soil bacteria. These bacteria invade the roots of legumes, form colonies, and so irritate the roots that nodules are formed around the colonies. Here the bacteria use part of the plant's sap for food, and carry on the nitrogen-fixing process. Part of this nitrogenous material is used by the plant, but the remainder is added to the soil when the plant dies and its roots decay. As you can see, legumes are able to obtain the nitrogen they need even where the soil is deficient in nitrogen compounds. In addition, they actually add to the supply of nitrates in the soil. As far as nitrogen is concerned, they leave the soil more fertile than before they were planted. For this reason, such legumes as clover, vetch, soybeans, peas and alfalfa are commonly used in crop rotation to increase the nitrogen content of the soil. The odd partnership between legumes and nitrogen-fixing bacteria is another example of *symbiosis,* in which two living things join together in a partnership from which both benefit.

In addition to the manufacture of nitrogenous plant food, legumes add to soil fertility in two other ways. They make the topsoil deeper and more fertile, and they help other plants extend their roots deeper into the ground. The roots of leguminous plants are able to penetrate hard subsoil, often to a depth of many feet. In the process, they break up the subsoil, enabling bacteria and water to penetrate more deeply into it. By following the roots of legumes, the roots of other crop plants are able to penetrate more deeply into the subsoil. This helps them reach the minerals that are located in great abundance in the deeper earth. Legumes also add humus to the soil if they are "plowed under" at the end of the growing season. Whenever a plant decays in the soil, the substances of which it was composed are returned to the soil. Crops that are plowed under add "green manure" to the soil—plant tissue that will decay and once again be available for use as plant food. Legumes make the best green manure because of the nitrogen content mentioned earlier, and because of the wandering roots that abstract minerals and plant food from the subsoil many feet below the surface.

How is tableware silver-plated?
Most of the silverware used in the average home is not solid silver, but iron with a thin coating of silver on the outside. In the manufacture of plated silverware, a piece of iron is first formed in the

shape of the article desired. A coating of silver is then applied to the article by an electrochemical process. Let's see how this is done.

First we must have a tank made of a material that will not conduct electricity—one of glass for example. The tank is then filled with a solution of silver nitrate. The iron articles and a strip of pure silver are hung in the solution, all completely covered. One terminal or post of a battery is then connected to the silver and the other post to the articles to be plated. The negative post is connected to the iron pieces and the positive post to the silver. In a short time, the silver bar begins to be eaten away, and a layer of silver builds up on the iron articles. To make the silver stick better, the partially plated articles are removed periodically and polished, after which another coating of silver is applied. This produces a thicker and stronger layer of silver that will last longer in service.

The process by which silver transfers to the iron article is called *electrolysis.* When silver nitrate is dissolved in water, it breaks up into two particles called *ions.* One of these ions is pure silver, and the other is the nitrate ion. The silver ion has a positive electrical charge while the nitrate ion has a negative charge. You will recall that we connected the negative post of the battery to the iron articles to be plated. This negative post attracts the positive silver ions which move to the iron pieces and attach themselves to the iron. They then give up their positive charge and become normal atoms of silver. At the other end of the tank, the bar of pure silver is connected to the positive post of the battery. This attracts the negative nitrate ions which attack the silver and form more silver ions. This process results in a continual movement of positive silver ions from the bar of silver to the iron articles. When they reach the iron, they lose their charge and adhere to the iron. In this way, a layer of pure silver is plated on the iron and silver-plated tableware is the result.

Many other metals can be plated on less expensive base metals in much the same way. Gold, copper, chromium, cadmium, palladium, nickel and many others are plated on other less expensive metals for improved appearance or performance. For most purposes, a plated article is just about as useful as one made of the solid metal. In some cases the plated product combines the good features of both metals —for example, the strength of steel and the corrosion resistance of nickel.

What are man-made elements?
For many years, scientists believed that there were only 92 chemical elements. Of these, several occurred in such small quantities in nature that they long defied discovery. By 1947, however, even the most elusive had been discovered or produced artificially. But by this time the use of new instruments and techniques had changed scientists' ideas about the number of chemical elements they might produce. There are 102 elements known at the present time and at least several more are predicted.

Element 93, named *neptunium* after the planet Neptune, was produced in 1940 by McMillan and Abelson. It was obtained by bombarding uranium-238 with subatomic particles. Later in the same year, G. T. Seaborg and others prepared element 94 which they named *plutonium* after the planet Pluto. In 1944, this same group of scientists prepared element 95, *americium*. It is obtained by bombarding an isotope of plutonium with neutrons. Other man-made elements include *curium, berkelium, californium, einsteinium, fermium, mendelevium,* and element No. 102, as yet unnamed.

All of the man-made elements have radioactive isotopes. Isotopes of an element, you will remember, have the same chemical properties but different weights, owing to a different number of neutrons in their nuclei. The half-life periods of these isotopes vary from a few minutes for an isotope of einsteinium to 2.2 million years for neptunium-237. Some of these elements are presently used in atomic reactors, but we can only conjecture upon the uses to which they may eventually be put.

How fast does human blood flow?
The speed at which blood flows through the body has been determined by measuring the time required for a dye injected into a vein in the neck to return to an artery in the same region. In such a journey, the injected dye must pass through the right auricle and ventricle of the heart, the lungs, back through the left auricle and ventricle, and out to the artery of the neck region. This journey takes about ten seconds! Equally astounding is the combined length of the blood vessels in the human body. Blood leaving the heart travels through arteries that get progressively smaller and smaller as they approach the various organs. Blood leaves the organs by means of veins that get progressively larger and larger as they approach the

heart. Between the smallest arteries and veins there exists an elaborate network of microscopic capillaries. These capillaries wind around the cells of the body providing food to the cells and obtaining waste products in return. It has been estimated that if the capillaries were removed and placed end to end in one continuous filament, the delicate microscopic capillary would measure 62,000 miles—long enough to extend two and one-half times around the earth!

Are red, blue and yellow the primary colors of light?

No. It's possible to form white light by combining rays of red, green and blue-violet light. In fact, any color at all can be obtained by combining light of these colors in the proper proportions. For these reasons, red, green and blue-violet are known as the *primary colors of light*. But from our dealings with paint and pigments most of us recall the primary colors to be red, yellow and blue. How can there be two sets of primary colors—one for light and another for pigments? Why is it that paints don't combine to produce the same colors as light? To understand the answers to these questions we must look into the nature of pigments. A certain paint is called yellow because the light that it reflects is *predominantly* yellow. But ordinary yellow paint is impure and also reflects a considerable amount of green light. Another type of paint is called blue because it reflects light that is principally blue in color. But blue paint also reflects a certain amount of green light. Thus we have two paints, blue and yellow; each reflects green light in addition to its own color; each absorbs all other colors. What happens when blue and yellow paint are mixed together? The blue pigment absorbs yellow light; the yellow pigment absorbs blue light; the only color that can be reflected is green! So blue and yellow paint can be combined to produce green because of the impurities present! If blue and yellow *light* are mixed, the resulting color is white, for blue and yellow are complementary colors of light. The fact that pigments do not behave like light can be explained completely on the basis of the impurities in most pigments and their selective absorption of various colors.

Why does a window pane behave like a mirror at night?

When a window pane is viewed at night from inside a lighted room, it acts as a mirror and we can see our image in it. This is because only part of the light that strikes the window is transmitted through the glass. A portion of it is reflected back into the room. It is this reflected light that is responsible for the mirror-like behavior of the window. During the daytime, the light entering the room from the outside is great because of the brightness of daylight. The transmission of this daylight through the window effectively masks the small amount of reflected light that could produce the mirror effect. At night, however, little light from outside can be transmitted through the window into the room. The reflected light then predominates and the window acts as a mirror.

Are more meteors seen after midnight than before?

Meteors are nothing more than small fragments of iron or rock which have entered the earth's atmosphere. Because of their high speed through the atmosphere, they're heated by friction and become hot enough to emit light and be seen. Studies have shown that meteors, or shooting stars if you prefer, are about twice as numerous after midnight as in the early hours after sunset. This is a result of the earth's rotation about its axis as it revolves around the sun. Meteoric fragments are dispersed to a greater or lesser extent throughout the space through which the earth must travel in its orbit. After midnight, we're on the leading side of the earth as it moves along its orbit and we see meteors arriving from all directions. In the hours before midnight, we are on the trailing side of the earth and see only those meteors which are moving fast enough to "catch up" with the earth. As a result, we see more meteors after midnight than before.

Why do some combinations of musical notes sound discordant?

Without attempting a rigorous definition, we can define music generally as a pleasing combination of sounds. Oddly enough, physicists tell us that notes which make harmonious combinations have frequencies (number of vibrations per second) that can be expressed *in the ratio of small whole numbers.* For example, the *octave* consists of two tones whose frequencies have the ratio 2:1. Another harmonious combination is the *fifth,* whose frequency ratio is 3:2. One of the most pleasing combinations of all is the *major chord* consisting of three tones having a frequency ratio of 4:5:6. An example may serve to show how this works. The C major chord is made

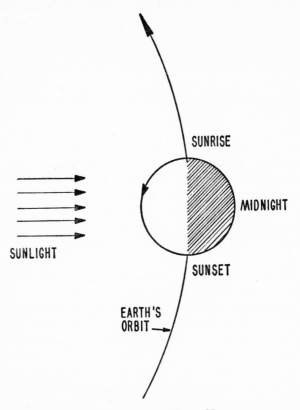

FIG. 27. WHY METEORS ARE MORE NUMEROUS AFTER
MIDNIGHT

During the hours of darkness before midnight, we see
only those meteors which are traveling fast enough to
catch up with the earth as it moves in its orbit. After
midnight, we are on the front side of the earth as it
moves along. We are then able to see meteors arriving
from all directions.

up of the notes C, E and G. The frequencies of these notes (in the
physical scale used by physicists) are 256, 320 and 384 vibrations per
second, respectively. If each of these frequencies is divided by 64,
the resulting numbers are 4, 5 and 6. These notes, therefore, are in
the ratio 4:5:6. In a similar manner, the *minor chords* are made up
of notes in the ratio 10:12:15. Now let's investigate a discordant

combination of notes. If F and E are played together too often, most musicians rapidly lose their audience. These notes are in the ratio 853:1600. The fact that F and E are not in the ratio of small whole numbers seems to make them sound discordant to the human ear.

What is the "sound barrier"?

The "sound barrier" refers to the stresses to which an airplane is subjected when flying at speeds very close to the speed of sound. In ordinary low-speed flight, the forward portions of a plane send out a pressure wave which travels at the speed of sound. This pressure wave results from the build-up of particles of air as the plane rushes forward. Naturally, these pressure waves travel faster than the airplane in ordinary subsonic flight. In doing so they seem to affect the air in such a way that it moves smoothly over the approaching wing surfaces. If the plane travels at the speed of sound, however, the air ahead receives no advance pressure wave. Instead, the pressure wave builds up in front of the wing since both wing and pressure wave are moving forward at the same speed. This results in a shock wave which induces relatively great stresses in the wing. In the days preceding supersonic flight the term "sound barrier" came into use to describe the increased stresses that were expected when a plane reached the speed of sound. Actually, the term has turned out to be a misnomer since the speed of sound has proved to be no barrier at all. The problem has been met and surmounted by aeronautical engineers in the design of supersonic airplanes.

Oddly enough, it's easier to crash the sound barrier at high altitudes than at altitudes near sea level. This is because the speed of sound decreases as the plane climbs to higher altitudes. The speed of sound is 760 miles per hour at sea level and only 660 miles per hour at 36,000 feet.

Perhaps you have heard a loud explosive-like sound and felt its vibration when fast aircraft are flying near by. Such occurrences are commonplace to those living near jet air bases. This loud sound, or sonic "boom," is caused by the shock wave produced by a high-speed plane as it passes through the sound barrier.

What do the smallest fish eat?

We might begin by assuming that fish eat smaller fish which, in turn, eat still smaller fish, and so forth, but sooner or later we're faced with this fact: Only plants containing chlorophyll are able to manufacture carbohydrate food from the raw materials of nature. Without plant life, all of the animals of the sea would eventually starve to death. The plants that compose the flora of the oceans and of our fresh-water bodies are called *algae*. Although they have little direct economic use, their indirect value to man is tremendous. They constitute the only basic source of carbohydrate food available to aquatic animal life.

The many forms of algae are commonly classified according to their color as blue-green, green, brown and red. Each type contains green chlorophyll, the basic food-manufacturing substance of nature, but in some types the color is masked by the presence of another color. As we have indicated, algae are predominantly aquatic forms of life.

The blue-green algae usually consist of a single cell, or a single string of cells. Their cell structure is similar to that of bacteria and is therefore much simpler than that of the higher forms of life. Since they contain chlorophyll, they are able to combine water and carbon dioxide into food in the presence of sunlight. They are some times responsible for a bad odor or disagreeable taste in drinking water, but otherwise they are of no direct economic interest. They can thrive under temperatures close to the boiling point and for this reason may represent a form of life that was prevalent when the earth was young and much hotter than it is today.

Green algae, like the blue-green, usually consist of a single cell, but they differ in being able to move about in the water. They are often equipped with microscopic whiplike arms, called *flagella*, which enable them to move about very rapidly. While most varieties live in fresh water, a few are common in the ocean. Most of the tropical forms secrete lime and assist the coral polyps in the formation of coral reefs. The fresh-water types are usually microscopic in size but some of them form colonies which can be seen with the unaided eye. The green scum on the surface of ponds and the "water bloom" of small bodies of water are a result of such colonies.

Brown algae usually live in the ocean and most varieties grow to considerable size. Some of the large kelps of the South Atlantic are algae that have grown to a length of a hundred feet or more. Another well-known form is *Sargassum* (gulfweed) which is usually found floating unattached. Huge collections of this material which have been brought together by the action of winds and ocean currents have given the name to the *Sargasso Sea*. One of the more curious forms of brown algae is the *diatom*. It consists of a single-celled form that literally surrounds itself with a glass house—a shell of silica, the mineral of which quartz is formed. Although they are microscopic in size, great quantities of them settle to the ocean floor in deposits of great thickness. Ancient deposits of the shells of diatoms, which have later been lifted above the sea, form our present-day deposits of *diatomaceous earth*. Because of its extreme fineness and hardness, this substance is used to produce the best polishing materials known to man.

Red algae are able to live in deeper water than any of the other types. This ability is due to their *fluorescence*. *Fluorescence* enables them to absorb light of one color and change it to light of another color. As we have mentioned, all algae contain chlorophyll which manufactures food in the presence of sunlight. Despite its green color, chlorophyll must have *red* light if it is to work properly. Unfortunately, red light is able to penetrate ocean water only to relatively shallow depths. The greater depths receive little except blue light. Red algae solve this problem by changing blue light to red! In this way, the chlorophyll in red algae is furnished light of the proper color even at the greater depths.

As you can see, algae provide a ready source of food for the smaller forms of sea life wherever they exist. For the most part, the smaller animals feed on the microscopic plants while the larger animals eat the smaller ones, and so on, ad infinitum. Other forms of sea life probably feed directly on the larger kinds of algae. All in all, the various types of algae provide the basic food products for all of the aquatic animals; they are truly "the pastures of the sea."

Why does ordinary glass sometimes crack under extreme temperature changes?

Glass cracks easily under extreme temperature changes because it's a very poor conductor of heat and because it expands quite a bit when heated. When a glass container is heated rapidly, there isn't sufficient time for all portions of the glass to heat evenly. The part subjected to heat expands to a greater extent than the part away from the heat. This uneven expansion causes great stresses within the glass and it eventually cracks. Pyrex glass expands only one-third as much as ordinary glass and is therefore less susceptible to the effects of temperature changes.

Why do some rivers have characteristic colors?
There are many "colored" rivers in the world: "white" rivers like the White Nile, the Red Rivers of the United States, the Yellow River of China, and the Blue Danube of Europe. Most of these "colored" rivers derive their reputation for color from material held in suspension by their water. Since this material will vary with the particular terrain through which the river flows, each river assumes a color peculiar to the region. In some cases, the land imparts chemicals to the water which combine to produce a startling effect. Some Algerian rivers are colored by just such a process. One tributary of a river contains iron and another gallic acid taken up from peat marshes. As the tributaries combine, the two ingredients mix to produce a river literally "as black as ink." This is to be expected as iron and gallic acid are used in ink-making.

Perhaps the most spectacular of them all is the Rio Tinto, in southern Spain. It starts out as emerald green, high in the hills near the old mines of Tarshish—the same mines that were worked in the days of King Solomon. The small stream flows through land that is covered with outcroppings of *iron pyrite*. This mineral provides *ferrous sulfate*, a brilliant green chemical. Consequently, the Rio Tinto and other streams in the area are brightly colored by the dissolved chemical. As the water flows down mountain valleys covered with wild lavender, the river seems to change color to purple-brown. When it runs through sandy stretches and collects behind sand bars, the resulting pools seem to turn red. Later, as the stream draws closer to the sea, the water changes completely to blood-red. This results from a chemical change by which the dissolved ferrous sulfate, which is green, changes to ferric oxide which is as red as rouge. During the course of its travels, the Rio Tinto changes through all the hues from bright green to crimson red. No wonder the people of that area have long called it Rio Tinto—the Colored River.

Why is water more effective than alcohol in preventing an automobile engine from overheating?
The car owner who forgets to replace his alcohol antifreeze with

water may regret his lapse of memory if the weather gets very hot. This is because water has a greater heat capacity than the alcohol that is often used as an antifreeze. Each ounce of water can remove up to 30 per cent more heat from the engine than can an alcohol-water mixture. Water is said to have a higher *specific heat* than alcohol.

The calorie is a popular unit of heat measurement. It's the quantity of heat required to raise the temperature of one gram of water one degree centigrade. The specific heat of a substance is the amount of heat required to raise one gram of *that substance* one degree centigrade. Thus, the specific heat of water is 1. It takes about 0.58 calorie to raise the temperature of one gram of alcohol one degree centigrade, so the specific heat of alcohol is 0.58. In other words, a given amount of alcohol requires less heat to accomplish a given temperature rise than does water. Let's assume that an automobile engine raises the temperature of the cooling liquid by 35° C. This means that the temperature in the radiator—the coolest point in the system—is 35° C. cooler than the liquid coming out of the engine. Under these conditions, a gram of water would absorb 35 calories (one for each degree) as it passes through the engine. Pure alcohol, on the other hand, would absorb only 20.3 calories (0.58 for each degree). As you can see, alcohol cannot absorb as much heat as can water. If the weather becomes hot, the cooling system using alcohol is more likely to overheat.

Specific heat is really the ratio between the heat capacity of a substance and the heat capacity of water. It so happens that water, along with its other unusual characteristics, has a greater heat capacity than most other substances. For this reason most specific heats are likely to be less than 1. The following table lists the specific heats of a few ordinary substances.

Water	1.00	Copper	0.09
Ice	0.53	Mercury	0.03
Soil	0.20	Marble	0.21
Aluminum	0.22	Glycerine	0.55
Lead	0.03	Alcohol	0.58
Glass	0.14	Platinum	0.03

The relatively high specific heat of water compared to soil plays an important role in moderating the climate of coastal land areas. Since water requires more heat to reach a given temperature, the oceans do not heat up as fast as the land. Oceans, therefore, are cooler in summer than adjacent land masses. This enables them to act as cooling influences on near-by land. During the winter season, just the opposite effect takes place. Oceans take longer than near-by land in the cooling-off process and are usually warmer than the coastal areas. In this way, the high specific heat of water keeps the seashore cooler in summer and warmer in winter.

How can so many different kinds of cheese be made from cow's milk? Cheese-making is an ancient art, and like the making of butter, it has developed into a widespread and important industry. In the present-day manufacture of most cheeses, cow's milk is warmed to encourage the growth of bacteria and to bring about coagulation by *rennet*. Rennet is a substance obtained from the stomach of cows. After a time a curd forms which is removed from the fluid, known as whey. It is then salted, pressed and placed in the curing room, where bacteria, molds and other microorganisms work to produce the texture, flavor and aroma of the desired kind of cheese. In order to be sure that the desired kind of cheese results, it's necessary to control the growth of various essential microorganisms. In grandfather's day, this was accomplished by varying the acidity of the milk, the coagulation time, the amount of salt added, the length of time for ripening, the temperature of the curing room and countless other environmental factors. This procedure determined which species of microbe would flourish at the various stages of the cheesemaking process. Such methods made cheese-making a highly skilled art, passed down from generation to generation with the secrecy of a confidential military report. With scientific methods, however, have come certain complications. The milk used in today's cheese factories is cleaner and pasteurized, and the cheesemaker can't rely on the chance presence of necessary microorganisms. He must use cultures of carefully identified organisms to be sure that his cheese turns out to be of the desired kind. It is now possible for him to buy cultures of organisms for producing Cheddar cheese, for producing Limburger cheese with its outstanding characteristics, for making the holes in Swiss cheese, and for producing the flavor and aroma of just about any cheese that the buying public may desire. Cheese-making, you might say, is merely the art of raising the "right" kinds of bacteria.

251

How did "horsepower" get its name?

During the latter part of the eighteenth century, horses were still used in England to pump water out of coal mines. At about 1780, James Watt tried to induce mine owners to substitute his newly developed steam engine for the horses. Typical businessmen, his prospective customers wanted to know how well his steam engine compared with horses. In order to compare the engine with horsepower, Watt measured the rate at which a strong draft horse could do work. He found out that a good horse could do work at the rate of 550 foot-pounds per second. This is equivalent to pulling a weight of 550 pounds through a vertical distance of one foot in one second. This has since become the standard unit of power and is called the *horsepower*. It was then a simple matter for Watt to measure the rate at which his engines could do work. He could make a direct comparison between his steam engine and horses.

The electrical unit of power is called the *watt* in honor of Watt's contribution to science. One horsepower is equal to 746 watts. Since a *kilowatt* equals 1,000 watts, one horsepower is equal to approximately $\frac{3}{4}$ kilowatt.

For relatively short periods of time, a man is able to work at a rate approaching $\frac{9}{10}$ horsepower, but his normal output for extended periods of time is more nearly $\frac{1}{8}$ horsepower. The small motors used about the home for vacuum cleaners and refrigerators are generally rated at a fraction of a horsepower, about the same as a man is.

How was the universe formed?

Two of the many theories concerning the formation of the universe are known as the *big-bang* and *continual creation* theories. Both attempt to account for the apparent expansion of the universe, a phenomenon by which every bit of matter in the universe seems to move away from every other bit at a speed which depends upon their distance of separation. The greater the distance of separation of two stars, for instance, the greater their observed speed of separation.

The big-bang hypothesis suggests that the universe came into existence as a result of a gigantic explosion. During this explosion all of the material of the universe was thrown out from a central point toward the reaches of outer space. The velocities of the many "fragments" of this explosion would then be expected to vary all the way from practically zero to the speed of light itself. Astronomers have even calculated the time in the dim past at which such an explosion would have had to take place. Imagine, for a moment, that time is running backward instead of forward. Then all of the celestial bodies would be *converging* instead of diverging. If this process were kept up long enough, every particle of matter in the universe would end up at some central location and a great lump of cosmic material would be the result. Since scientists know the location and speed of many of the stars, it's relatively simple to calculate the length of time necessary for them to arrive at a central point, if their directions were reversed. Such calculations indicate that the big-bang would have occurred six or seven billion years ago. The greatest drawback to this theory is a subtle one. A few billion years from now, all of the stars will have moved very far apart. If the big-bang theory is valid, it will be impossible for people on earth to see any but a few of the galaxies that happen to be close to the sun and traveling at about the same speed and direction. This would mean that we see so many galaxies now simply because we happen to live at an early stage of the big-bang expansion—at a time before most of the galaxies have had time to disappear from view forever. Many astronomers find it hard to believe that we are lucky enough to live in such a particular period of time when so many other less favorable periods are possible.

The continual creation hypothesis eliminates this objection to the big-bang theory by introducing the idea that matter is being continually created *from nothing,* everywhere in the universe! This creation of matter is supposed to be going on at a rate that would just equal the loss of matter as galaxies disappear over the visible horizon. It is supposed that all space has the property of producing matter in such quantities as to balance the loss to our observable universe by its expansion. Scientists know the rate at which galaxies are disappearing, and they know the amount of matter that each such disappearance represents. This makes it an easy matter (for them) to calculate the rate, on the average, at which matter must be created in order to preserve a constant amount in the observable universe. This rate is extremely minute because of the tremendous volume that our universe takes up. In terms of everyday things, the theory can be satisfied if one hydrogen atom is created in the average-size room every hundred thousand years! This rate is so small that it would hardly be recognized even if the hydrogen atom happened to pop out of this page at this instant. As fantastic as this theory may sound, it has received acceptance and support by many experts. Of course, no theory is ever accepted completely—it receives continual study and serves as a guide to the nature of things until a better theory is found.

Does the expression, "pouring oil on troubled waters," have any basis in scientific fact?
The expression, "pouring oil on troubled waters," is generally used to refer to something that soothes a troubled situation. However, it does have a basis in scientific fact. When oil is poured on rough water, the area covered seems to be more calm than the surrounding, uncovered sea. This is one of the factors which makes it possible to see an "oil slick" on the ocean. Oil accomplishes this feat by providing a surface film which is stronger than that of water alone. This makes the oil-covered area less susceptible to the effects of ocean breezes.

What do whales have to do with perfume?
Ambergris is a solid, fatty, waxlike substance produced in the intestines of the sperm whale (cachalot). It is used in the perfume industry as a fixative to make essences retain their fragrance. No completely satisfactory substitute has been found for the substance and it's therefore quite valuable, being second only to choice pearls in the value by weight of products obtained from the sea. It is believed to be connected in some way with the beaks of cuttlefish and squid that are often found in it. Cuttlefish and squid are the chief food of the sperm whale. A theory holds that ambergris is found only in sick whales as a secretion brought about by the illness. It has a variegated appearance ranging in color from light to dark gray. In the crude state it gives off an agreeable fragrance which faintly resembles that of sealing wax. It was formerly used as a medicine, but this use died out as its lack of efficacy became known. It was also used as an incense in the Orient, but its value now limits its use to the perfume industry. It is a rare substance, and is not usually taken from whales directly. It is most commonly found by accident floating on the ocean, or cast up on a beach. Occasionally, large pieces weighing as much as 150 pounds are found. Such a piece may be worth as

much as $25,000 in the crude state. When properly processed, its value might be several times that amount. The synthetic production of ambergris or a substitute would seem to be a profitable area of work for organic chemists.

What are "Brownian movements"?

One of the most convincing proofs of the existence of molecules and their state of continual motion was discovered in 1827 by an English scientist, Robert Brown. The molecules themselves, of course, are much too small to be observed directly, and indirect methods must be used to study their motion. Brown, a botanist, noticed that very fine particles suspended in a liquid move about in an irregular and lifelike manner. He attributed this motion to uneven and random bombardment of the particles by moving molecules of the liquid. Subsequent research and mathematical study have amply borne out his hypothesis. These erratic motions are called *Brownian movements* and can be observed easily with the aid of a microscope. All one has to do is suspend some very fine particles in water and view them under a high-power microscope under proper illumination. Brownian movements can also be observed in fine dust particles suspended in air, as well as in many types of particles suspended in liquids.

Has it ever rained fish?

Unbelievable as it sounds, a rain of fish did actually occur in 1817, at Appin, in Scotland. It consisted of a downpour of small herrings, a feat that nature repeated in 1830, at Islay, in Argyllshire. Some sixty years ago there was a shower of small frogs in the west of England and in 1900 a thunderstorm brought down more of the creatures near Liverpool. Even this doesn't exhaust the marvels of nature, for many other curious effects have been connected with rainfall. For example, there was a shower of red rain in 1608 at Aix, during which large red drops of liquid were seen on the cemetery and the walls of the church. Needless to say, this "shower of blood" was not taken lightly by the frightened inhabitants. Red rain has been recorded many times since then, for instance at Vienna and in Italy in 1901, in Cornwall and Hamburg in 1902, and in England in 1903. The explanation probably lies in the fact that large quantities of *algae* were brought down by the rain. Algae are tiny plants measuring less than one-thousandth of an inch in diameter—the simplest forms of vegetable life.

Black rain is another oddity that has visited the British Isles. In 1862, four showers of black rain fell in Scotland. They were probably the result of volcanic dust brought to earth from the higher atmosphere. Yellow rain has also been recorded and pollen is suspected of being the coloring agent.

While such curiosities of nature are startling, they all have natural explanations. The herrings—and small ones at that—were probably picked up by a waterspout at sea. The frogs probably enjoyed a similar experience as a result of a whirlwind, either from a swamp or from a meadow. In any event, no rainstorms of fish or frogs have been recorded far from either seacoasts or swampland.

How is altitude measured?

Barometers are devices used to measure atmospheric pressure. The ocean of air we live in presses down on the earth and everything in it. The amount of this pressure, at sea level, is 14.7 pounds per square inch. It results entirely from the weight of the air above the earth. Slight variations in atmospheric pressure (a few per cent) are known to be related to the weather and are useful in weather forecasting. But atmospheric pressure varies greatly with height above sea level and this fact makes it possible for airplane pilots to determine their altitude. Naturally, the pressure of the atmosphere is less on a mountaintop than at sea level because there is less air above the observer. If a person were to go high enough, perhaps five or six hundred miles high, there would be practically no air left and the atmospheric pressure would be almost zero. This makes it possible to measure the altitude of any particular location by means of a barometer. The pressure of the atmosphere drops about 1 pound per square inch (for elevations up to a few miles) for every 1,800 feet of ascent. Let's assume that the atmospheric pressure at sea level is 14.5 pounds per square inch. (It will vary a few per cent from day to day.) Let's further assume that a barometer on a near-by airplane records 12.5 pounds per square inch. The difference between these readings is 2.0, so that the plane must be 3,600 feet high (1,800 x 2.0).

The most convenient type of barometer for this purpose is the *aneroid barometer*. It consists of a small sealed metal box from which most of the air has been removed. In addition, it has a corrugated face that moves in and out as the atmospheric pressure varies. This slight movement is magnified by a system of levers and is used to move a pointer. The scale can be graduated in pounds per square inch, inches of mercury or, in the case of altimeters, directly in feet.

How did rubber get its name?

One of the sights that greeted Columbus on his visit to the New World was that of native children playing with balls that bounced. Investigation disclosed that these intriguing balls were masses of gum from a local tree. He found them so fascinating that he took a few back to Europe. Three hundred years later, Joseph Priestley, the discoverer of oxygen, found that rubber could be used to rub out pencil marks and he gave the substance its name. Macintosh, a Scotsman, was the first to make a rubberized raincoat, and Dunlop, an Englishman, made the first rubber tire by wrapping a rubber tube around a wheel. But products made in this way are much too stiff in winter and much too sticky in summer. A solution to the problem was discovered by accident, in 1839, by Charles Goodyear who let a mixture of rubber and sulfur spill over the hot kitchen stove. When he tested the "ruined" sample, he discovered that rubber, when heated in this way with sulfur, stretches and snaps and has the desirable properties that we expect in rubber today. Rubber of this kind is said to be *vulcanized*. After a great deal of work with rubber, Goodyear had found success by accident. But he knew the value of the material that chance had produced for him and he followed through with his discovery. As Pasteur once so wisely stated, "chance only favors the mind that is prepared."

Why does a shower curtain move inward when the shower is running?

If you like to take showers, you've probably been annoyed at one time or another by the tendency of a shower curtain to move toward you with persistent regularity. The main reason for this annoyance is the speed at which the water moves within the shower. The rain of drops drags near-by air along with it causing circulating air currents to flow within the enclosure. When the shower is running hard, these currents of air get moving quite fast. We then have a condition in which the air within the shower enclosure is moving considerably faster than the air in the other part of the room. In other words, the air on one side of the curtain is moving faster than the air on the other side. Whenever such a situation exists, there is a force tending to push the curtain toward the region of high-speed air. The same sort of thing is responsible for the "lift" that keeps an airplane from falling to earth. The wing surfaces are designed in such a way that air must move faster across the top surface of the wing than across the bottom. Since the high-speed air is above the wing, there is a force pushing the wing upward.

These examples illustrate Bernoulli's principle, which states that the pressure in a fluid (a liquid or gas) decreases as its velocity increases. Let's see how this can be applied to the shower curtain problem. In the usual situation, air on the inside surface of the curtain is moving at some reasonably high speed. Air on the other side is presumably at rest, or at best, moving slightly. This means that the pressure exerted on the inside surface will be less (because of Bernoulli's principle) than that exerted on the outside surface. This results in an unbalanced force which pushes the curtain toward the faster-moving air.

If you care to take the time, you can prove the truth of this principle by blowing across the top surface of a piece of paper. Hold the piece of paper by the two corners closest to you. The portion you are holding will be horizontal but the far end will droop because of its weight. Now bring the edge that you are holding up and touch your chin with it. If you now blow across the top of the paper, the paper will rise! You will probably have to experiment a bit before you discover the best direction in which to blow. But once you have acquired the knack, you will have no difficulty in convincing yourself of the truth of Bernoulli's principle.

Before we leave the subject, we ought to look into the fact that chimneys draw better on windy days. As you may have guessed, this is also a result of Bernoulli's principle. On windy days, air whistles past the chimney top at a high rate of speed. This means that the air pressure at the top of the chimney is *lower* than usual. In any event, it's lower than the air within the living room which is just about stationary. Since the pressure in the house is greater than the pressure at the chimney top, there is an increased air flow in the chimney. The greater pressure within the house forces air out the chimney at a higher than normal rate.

Why does the moon seem to follow us as our auto speeds along the road?

Have you ever heard a youngster ask why the moon (or sun, for that matter) seems to follow a moving automobile? I remember marveling as a youngster at its apparent motion; no matter how fast the automobile went, the moon followed right along. Of course, this apparent interest of the moon in our travels is completely a psychological reaction on our part. As we speed along the road, it's only natural to expect the countryside to fly past in the opposite direction. The moon, after all, is merely a part of that countryside and we subconsciously expect it to behave in the same manner as trees, houses and every other fixed part of the scenery. If the moon were to fly backward with the passing scene, everything would be "normal." What really happens is quite simple. The moon's distance from the earth is extremely great compared to the distance that our automobile moves in a few minutes. This means that the angle at which we see the moon does not change perceptibly as our auto moves along the highway. If our auto travels a straight path, the moon will maintain essentially the same angle with respect to the observer. The angle of everything else, however, changes rapidly as the objects involved fly backward. Since the moon's direction changes much more slowly than the direction of near-by objects, we get the impression that the moon is moving right along with us.

The same sort of thing happens when the road is located in such a way that both near and distant objects are visible from the car. If intervening objects are hidden from view, the illusion is even more striking. Under these conditions, the distant objects seem to move with the car, although not quite as fast—as though trying to catch up. It all results from the fact that the direction of a distant object changes very slowly, while the direction of a near-by object changes very rapidly.

Why does the sun seem larger nearer the horizon?

The apparent increase in size of the sun and moon when they are near the horizon is entirely psychological. The human eye constantly estimates the size of objects near the ground by comparing them with other objects within view. When heavenly bodies are overhead, and out of the way of other objects, they look smaller to us. You can prove this to yourself at the next sunset by holding a piece of paper so as to shut out the view of other objects on the earth. The sun will then look no larger than when it is overhead. As far as the moon is concerned, it ought to look larger when it is directly overhead since it is then about four thousand miles closer to the earth than when it is at the horizon. This would make it seem about one-sixtieth larger. This difference is much too small, however, to be distinguished by the casual observer, and is completely masked by the psychological factors discussed above.

What shape does an egg have?

Although the word "egg-shaped" has become part of our language, there is really no general shape for the eggs of birds. The term "egg-shaped" refers to the shape of the eggs of our domestic fowl. Since these are descended from jungle fowl, they are really a kind of game bird, and the eggs of such birds are typically ovoid in shape. It's often possible to determine the kind of bird that laid an egg by a combination of its color and shape. The eggs of snipe and plovers, for example, are pearl-shaped, while the eggs of the grebes and swallows are biconical, having each end pointed. The night hawk and whippoorwill have oval-shaped eggs, while those of the owls and kingfishers are quite spherical.

Contrary to popular belief, it isn't always possible to determine the size of the bird from the size of the egg. Egg size doesn't necessarily correspond with the size of the bird that laid it. The egg of the guillemot of the cliffs, for instance, is very large in comparison with the size of the bird. Nor are all eggshells similar in appearance. Although most birds lay smooth eggs, those of the woodpeckers are quite glossy. The ducks, on the other hand, lay greasy eggs that can withstand the wet bodies of ducks coming from the water.

What is color?

You've probably heard the story of the man who, when asked the time of day, told his questioner how to make a clock. Describing the nature of color makes one feel like that man. We are fortunate indeed when we can find easy answers for simple phenomena, but the truth of the matter is that color just isn't simple. Scientists tell us that color is a sensation of the human brain, brought about when the eye receives light of a certain frequency. In order, then, to fully appreciate color, we must look into the connection between *frequency* and *light*. You've probably connected the term "frequency" with your radio receiver. By rotating the tuning dial of your radio set, you can change its frequency so that it "picks up" the desired station. The air about us is full of radio waves which could be picked up, but the radio set selects the ones of the *desired* frequency. This then is the essential difference between broadcasts from the different radio stations: they all have different frequencies, and this difference is all that is required to enable them to be separated by the radio set. But what is frequency? And more specifically, what are radio waves? And what does either have to do with light and color? Perhaps it will suffice to say that radio waves are merely energy in motion. The broadcasting station sends out a continuous flow of energy, some of which is picked up by your radio receiver. The motion imparted to this energy is of two kinds: the first, of course, is the motion outward from the station in all directions so that some of it can reach the various radio sets in the area; the second is a vibratory motion at right angles to the outward motion described above. It is this vibratory motion that is connected with frequency. The faster the vibrations, the higher the frequency. Our ordinary radio sets are tuned to select radio waves vibrating at the desired rate. If we design a radio set for higher and higher frequencies, strange things begin to happen. Instead of standard broadcasts, we begin to pick up short-wave broadcasts. As we go higher and add a picture tube, we pick up television broadcasts. Then we go higher still and pick up the strange transmissions of radar sets. If we could design a radio set for still greatly higher frequencies, *we would pick up light!* Yes—light is merely an extension of the same sort of phenomenon as radio waves! And light waves have a frequency just as do radio waves used to broadcast radio, television and radar transmissions. Scientists speak of this entire group of frequencies as the *electromagnetic spectrum,* or the spectrum, for short. Ordinary radio stations operate on a frequency of about one million vibrations per second. Light vibrates at frequencies centered about six million million vibrations per second.

Our eye acts somewhat like a radio receiver in that it can tell the difference between light rays of different frequency. Violet light has the highest frequency, about 7.5 (omitting the 12 zeros), and the colors follow through blue, green, yellow, orange and finally to red which has a frequency of about 4.0. Of course, there are no definite points where one frequency stops and another begins; there is rather a continuous blending from one color to the next containing all of the possible hues. When light enters our eye, it strikes a number of cone-shaped elements at the back of the eye. These cones are somehow able to respond differently to light of different frequency. The cones are of three types: those sensitive to red, green and blue-violet. If red light (at a frequency of 4.0) reaches the eye, the "red" cones are stimulated, and the brain produces the sensation of "red." Similarly, green light (5.0) stimulates the green-sensitive cones, and the same sort of thing happens for blue-violet. These three are called the primary colors of light. If red and green light reach the eye simultaneously, both the red and green cones are stimulated. The brain receives impulses from each set and interprets the color as a combination of the two, and we "see" yellow. In this way, it is possible to produce the sensation of all of the colors of the rainbow. In the case of sunlight, the brain interprets a mixture of all colors as white light, if they are in the correct proportions.

What causes clouds?
Clouds are nothing more than accumulations of mist, high above the earth. The amount of moisture that can be held as vapor in the atmosphere depends on the temperature, warm air holding more water vapor than cold air. Water vapor is a gas, of course, and therefore invisible, but if the temperature falls below a critical point, the molecules draw together and form minute droplets of water. This process, called *condensation,* is just the opposite of evaporation. The atmosphere is said to be *saturated* when it is carrying so much water vapor that only a slight reduction in temperature will cause condensation. The production of clouds depends upon that peculiar characteristic.

Air at 32° F. is saturated when it contains an amount of water vapor equal to $\frac{1}{160}$th of its own weight. Scientists have found that the amount of vapor necessary for saturation doubles for every 27° F. increase in temperature. This means that saturated air at 59° F. (32° + 27°) contains $\frac{1}{80}$th of its own weight of water vapor. Similarly, saturated air at 86° F. contains $\frac{1}{40}$ of its own weight of water vapor. As you can see, there may be more water vapor in the atmosphere during the summer than the winter because the temperature is higher. The degree to which the atmosphere is saturated with water vapor is called its *relative humidity*.

When saturated air is cooled, water vapor condenses even if the temperature is lowered only slightly. The gaseous water molecules come together to form a mass of innumerable droplets which we call a cloud. This usually occurs when saturated warm air rises to a higher altitude where the temperature is lower. Should the cloud of water droplets come in contact with a mass of warm dry air, they evaporate and the cloud disappears. This explains why the clouds are always changing form. As water changes back and forth from liquid to gas (vapor), the outline and form of the cloud undergo corresponding changes. This results not only from alternating currents of warm and cold air, but also from the weight of the droplets themselves, which causes them to sink slowly to lower levels. As gravity acts on the droplets, they eventually reach a layer of air that is warm enough to evaporate them.

One of the most unusual cloud formations known is the "cloud banner" that clings tenaciously to the top of the Matterhorn for long periods of time. The cloud almost resembles the smoke leaving a factory chimney. Actually, the "cloud banner" evaporates steadily at its extremity while being renewed at the same rate at its other end. Here's how it happens. When the sun shines on the side of the mountain, the rocks become hot even at high altitudes. Air in contact with the rock is heated, absorbs moisture from the mountain, and rises because it is warmer than the surrounding air. This process continues until the heated air passes above the top of the mountain. It then cools because it has lost contact with the warm rock of the mountain. Sooner or later it reaches an altitude that is cold enough to cause condensation and the "cloud banner" receives another mass of droplets. In this way, a column of air rising up the sunny side of a mountain forms a cloud as soon as it rises above the peak. Other examples of this type of cloud formation are found at the Cape of Good Hope where Table Mountain causes "the spreading of the tablecloth," and at Gibraltar where the "Plume" forms over the rock whenever a Levanter is blowing from the east.

Before we leave the subject of clouds, perhaps we ought to explain why some clouds cause rain while others do not. As we have seen, a cloud consists of tiny droplets of moisture which fall very slowly because of their small size. Most of the time, these falling droplets reach a warm dry layer of air and evaporate. Let's imagine for a moment that the air beneath a cloud is moist instead of dry. Under these conditions, the droplets grow in size as additional condensation takes place. If this process continues long enough, each droplet becomes a drop and falls to earth as rain. The ultimate size of the drop depends upon the moisture content of the air through which it falls, a fact which accounts for the different sizes of raindrops.

Why are mountaintops colder than their bases?

Casual thought might lead us to expect that the tops of mountains, being closer to the sun, would absorb more heat and become hotter than their bases. This doesn't happen in real life, of course. Many high peaks are covered with snow the year round, attesting to the continual cold at such heights. In order to fully grasp what is happening, we must understand how heat, which is really a form of energy, is transmitted from one point to another. Energy from the sun reaches the earth by means of radiation. This is the same process by which your body is warmed by a fire, even though the intervening air is cold. Energy, in the form of infrared rays, travels from one place to another, transporting heat energy in the process. The warmth you feel near a fire, or a hot iron, or any other hot body is due primarily to infrared rays. All hot bodies give off such rays. In addition to infrared rays, of course, the sun also gives off energy in the form of light rays. These are entirely similar to infrared rays except for the fact that we can see them. We could see infrared rays too if our eyes were modified slightly. This isn't as astounding as it sounds when we consider that infrared photography is an established fact. Cameras using infrared-sensitive film have been used to take pictures in a room "illuminated" only by a hot flatiron. This is pointed out to emphasize the fact that infrared (heat) rays and light rays are really versions of the same kind of radiation.

Getting back to our mountaintop, scientists tell us that our atmosphere allows visible light to reach the earth with little difficulty. Infrared (heat) rays, however, find it almost impossible to complete the entire trip through our atmosphere to the earth. Since light rays do get through, and since light rays are a form of energy, they are absorbed by the rocks and soil of the earth. This accumulation of energy in the earth's surface causes it to warm up to about 20° or 30° C. As with any warm body, the earth begins to give off infrared rays in order to lose some of this heat. But you will recall that infrared rays are not able to penetrate very far through the atmosphere. For this reason, they cannot escape readily into space. Our atmosphere reflects them back toward the earth. You might say that the atmosphere acts as a peculiar kind of blanket, allowing light energy to enter, but refusing to allow heat energy to escape. Conditions at the top of a lofty mountain are another matter, however. Since the air at such altitudes is very thin, infrared rays encounter no such difficulty and escape easily into outer space. This results in progressively cooler temperatures as the altitude increases and the atmosphere gets less dense.

The absence of an atmosphere on the moon causes the same sort of thing to happen there. Light and infrared rays heat the moon's surface to a very high temperature during the lunar day. At night, the radiation of infrared rays soon dissipates most of the surface heat and the temperature gets much colder than any place on earth.

What are antitoxins?

The child that suffers from diphtheria is ill, not so much from the throat inflammation produced by the bacteria, as from poisonous substances, known as *toxins*, that pass out of the bacteria. These toxins are molecular in size, and are readily dissolved in the blood. They pass throughout the body, causing injury to many tissues in the course of their travels. Recovery from a disease such as diphtheria will depend upon the production by the body of specific antibodies, or *antitoxins*. These substances "neutralize" the toxins, and minimize their effect on body tissue.

This toxin-antitoxin relationship can best be understood by means of an example. Let's imagine that we perform the following experiments on three guinea pigs. If we grow a batch of diphtheria bacilli in a suitable broth, we can pass it through a fine porcelain filter and obtain a sterile, bacteria-free filtrate. This filtrate will consist of water and the molecular-sized particles of the diphtheria toxin. The bacteria and the larger particles of bacteria-food will be filtered out. When a small portion of this toxin is injected under the skin of a healthy guinea pig, the animal will contract the disease and die in about a day or so. Even though no bacteria were injected into the guinea pig, the animal succumbs to the effects of the toxin.

Now, instead of a fatal dose, let's inject a second guinea pig with a series of inoculations, starting with an extremely weak dose. We will follow this up with several others of increasing strength over a period of several days. Eventually, we will inject the animal with a dose strong enough to kill a hundred untreated guinea pigs. But the animal will not die. The series of toxin injections will have stimulated the production of more and more antitoxin until the guinea pig has accumulated enough to neutralize the effects of even this last lethal dose. Our guinea pig No. 2 has become immune to the toxin of the diphtheria bacillus. His blood stream is rich in diphtheria antitoxin.

Finally, let's perform a third experiment using the clear serum obtained from the blood of our immunized guinea pig. If we mix a large amount of this serum with a fatal dose of the toxin, we find that the clear mixture soon becomes cloudy, indicating that a chemical reaction has taken place. If this mixture is injected into the third guinea pig, no ill effects will be observed. The mixing of toxin and antitoxin (serum from guinea pig No. 2) results in a neutralized solution that is not capable of poisoning the guinea pig. The third guinea pig is saved by the serum of the second. The same sort of thing takes place in the animal's body if we inject serum and toxin simultaneously.

Unfortunately, it is not possible to cure all bacterial diseases in this way because only a few disease microorganisms produce toxins that pass through the cell walls of the bacteria. These include those of diphtheria, tetanus, gas gangrene, scarlet fever and a few others. The majority of disease bacteria produce toxins that do not diffuse through the microbic cell wall but remain within the microorganism itself. These toxins can get into the blood stream only upon destruction of the microorganism. Strangely enough, it seems that such organisms injure the host by dying rather than living! As in the case of antitoxins, the body also produces antibodies to fight off such invading microbes, but the substance produced works directly on the microbe. Such antibodies are usually referred to as antibacterial antibodies to distinguish them from the antitoxins.

Is oxygen necessary for combustion?
Generally speaking, the term *combustion* is used to describe the chemical union of two or more substances accompanied by the production of light and heat. When coal burns in air, for instance,

carbon unites with oxygen, and light and heat are given off. Combustion isn't limited, however, to burning in air or oxygen; a jet of hydrogen burns quietly in chlorine gas, copper burns in vaporized sulfur, and phosphorus burns in gaseous bromine. Chlorine is particularly fond of supporting its own brand of combustion. It's a pale-green gas at room temperature, about 2½ times as dense as air. It's irritating to the throat and lungs and may be fatal if taken in large doses. If a wax candle is ignited and placed in a jar of gaseous chlorine, it burns with a yellow sooty flame. Such a candle consists of the elements carbon and hydrogen. In the burning process, chlorine removes the hydrogen from the wax, forming the compound hydrogen chloride. Black carbon is left over in the form of soot. Turpentine and many other compounds of hydrogen and carbon behave the same way. Many metals will glow or burn in chlorine. A thin sheet of warm copper will become red-hot in the gas, while metallic sodium reacts so violently with chlorine that it can be observed safely only from a distance. Pieces of arsenic or antimony burn spontaneously in the gas, forming a white smoke. When hydrogen and chlorine are mixed at ordinary temperatures, no perceptible reaction takes place—if it is done in the dark. But if the mixture is exposed to sunlight, the two elements combine with explosive violence! While we normally associate burning with oxygen, this substance is really only one of several that seem to satisfy nature's requirements for burning.

Does the harvest moon give more light than normal?
It is a fact that there is more moonlight in autumn than at any other time of the year. This is a result of a phenomenon known as the "harvest moon." During most of the year, the moon rises about forty-five minutes later each night. This means that in the course of five days, the moon will have delayed its time of rising nearly four hours (i.e., 45 minutes x 5 days). During the harvest season, however, the moon rises only a few minutes later each night so that it rises at nearly the same time for quite a few successive nights. This increase in the duration of moonlight reaches a maximum during the season of the harvest moon since at that time it rises about sunset for a number of successive evenings. It sheds light on the earth for most of the night, thereby allowing the farmers more time to work, should they be so inclined. The harvest moon is always the

full moon that falls nearest to the autumnal equinox, September 23. It is often seen just after sunset as an orange ball hanging low in the sky.

How does a hydrometer measure the amount of antifreeze in your car?

We are familiar with the fact that mercury is more dense than water, that oil is less dense than water. But it's often useful to know *how much* more dense one substance is than another. To make our

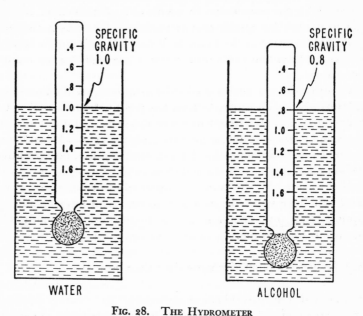

FIG. 28.   THE HYDROMETER

The hydrometer is a device for measuring the specific gravity, or relative density, of a liquid.

comparisons meaningful, however, we need a standard, and water has been generally used for this purpose. The density of a substance is its weight per unit volume. The density of water is 62.5 pounds per cubic foot. The density of alcohol is 49.5 pounds per cubic foot. If we divide the density of alcohol (49.5) by the density of water (62.5) we find that alcohol is 0.79 times as dense as water. This ratio, 0.79, is called the *specific gravity* of alcohol. The specific gravity of any substance is the density of that substance divided by the density of water. It's a useful tool in describing how dense, or heavy, a given substance is in comparison to water.

When we use a hydrometer, we are really measuring the specific gravity of a liquid. Fortunately, the substances normally mixed with water for antifreeze purposes have a density that is different from water. The specific gravity of the mixture will depend, therefore, on the amount of each liquid in the mixture. If alcohol is the antifreeze liquid, the specific gravity will vary from 0.79 for pure alcohol to 1.0 for pure water. By means of a suitable chart, it is possible to convert the specific gravity reading to a corresponding ratio of alcohol to water in the radiator. From this ratio it is possible to determine the amount of alcohol present, and the freezing point of the mixture.

A hydrometer is a hollow glass tube weighted at one end so that it will float upright in water. As with any floating body, it will sink to such a level that the weight of water displaced is just equal to its own weight. But suppose we float the hydrometer in a liquid that is less dense than water, alcohol for example. Obviously, the hydrometer must sink deeper into alcohol than water since a given volume of alcohol weighs less than water. In order to displace its own weight of alcohol, the hydrometer must sink deeper and displace more of the liquid. By proper calibration of the hydrometer, it is possible to measure the specific gravity of an unknown substance by noting the depth to which the hydrometer sinks. It will sink deeper in alcohol than in water; it will float higher in battery acid than in water.

Do ostriches bury their heads in the sand?

The story of the stupid ostrich that buries his head in the sand to avoid his enemies is pure fancy, a fable that is as untrue as it is ancient. It has become so widespread that it's often used as a convenient figure of speech to describe human self-deception. It's probably more nearly true in the latter application than in the one for which it was originally intended. Actually, the ostrich is not stupid at all, as birds go. After the chicks are hatched, they usually follow the parents around in search of food. The male, being a good father, sometimes tries to decoy an intruder; in so doing, he circles around the flock with outstretched wings and falls helpless, as though mortally wounded. In his own way, the ostrich "plays possum." It's quite possible that an over-ingenious father ostrich may have started the disparaging rumor by trying to draw attention from his family. At any rate, ostrich hunters tell us that the birds are anything but stupid.

In addition to playing possum, the ostrich enjoys another distinction. He is the giant among modern birds, reaching a height of eight feet and a weight of three hundred pounds. The male is black with white plumes; the female is gray-brown, providing excellent camouflage when the bird is lying on sandy ground. Ostriches, because of their great size, are not able to fly, but they can reach a speed of twenty-five miles an hour along the ground. They run with wings outstretched, with a stride of twenty-five feet.

The cock is polygamous, maintaining a harem of three or four hens which lay eggs in the same nest. He sits on the eggs at night in a nest consisting merely of a hole in the ground. During the day, he guards the nest or, weather permitting, kicks a little sand over the eggs and joins his wives in their search for food. This, to an ostrich, consists of just about anything that can be conveniently swallowed.

Is lightning good for anything?

One of the unexpected by-products of thunderstorms is the enrichment of the soil with nitrogen compounds. The United States Weather Bureau estimates that twelve pounds of nitrogen per acre fall to the earth each year as a result of lightning. This amounts to over 770 million tons for the entire earth! During the lightning flash, some of the energy of the electrical discharge causes atmospheric nitrogen and oxygen to unite forming the compound, nitric oxide. This compound consists of one atom of nitrogen and one of oxygen. It soon picks up an additional oxygen atom, however, becoming nitrogen dioxide. This new compound dissolves in rain water and falls to the earth in the form of a bath of dilute nitric acid. Chemicals in the earth then unite with this acid to produce calcium nitrate, an excellent plant food. So, if you're not particularly fond of lightning, you may derive some satisfaction from the fact that thunderstorms are good for the farmer in more ways than one; they give the earth the water it needs, and they help produce the fertilizer needed by growing plants.

What are the hottest and coldest spots on earth?

The world's highest temperature was recorded on September 13, 1922, at Azazia, Tripoli near the twenty-fifth parallel of north latitude. On that day, Azazia reached a temperature of 136.4° F. The highest temperature ever reached in the Western Hemisphere was 134° F. in July, 1913, at Death Valley, California. Death Valley is located at 36° N. latitude, more than a third of the way from the Equator to the North Pole, but it lies 264 feet below sea level; it comprises a deep valley walled in by a circle of mountains and it usually gets extremely hot in summer.

The coldest place on the earth, with the possible exception of the Antarctic mountaintops, seems to be Verkhoyansk in Eastern Siberia, just outside the Arctic Circle. In February, 1892, this spot recorded a frigid −94° F.! This compares with −76° F., the coldest polar temperature yet recorded. The average January temperature at Verkhoyansk is −56° F. and the record "high" for the month rises all the way to −13° F.! When people first settled this region, they found that the walls of their houses fell down. This was caused by the heat of their stoves which thawed out the ever frozen earth beneath their floors. Today, the floors are raised aboveground, thereby insulating the frozen ground from the warmth of their houses. It's probable that the tops of Antarctic mountains get even colder than Verkhoyansk because of their elevation. On the whole, the Antarctic is about 20 degrees colder than the Arctic. This is easily understood when we remember that the North Pole is covered by the sea while the Antarctic Continent rises at the South Pole to a high ice-covered surface thousands of feet above sea level. In the Arctic, birds, mammals and marine life extend as far north as there is open water. The interior of the Antarctic Continent, however, is almost entirely without life. Except for rugged lichens, life can exist only on the borders of Antarctica, where penguins, seals and whales may be found in fairly abundant quantities.

How does a flash bulb work?

The ordinary photographer's flash blub is filled with pure oxygen and aluminum or magnesium metal foil. It also includes a filament that gets hot when an electric current passes through it. This heat is sufficient to ignite the thin metallic foil, causing it to burn brightly in the presence of pure oxygen. When the batteries used in a photographic flash attachment get old, the current is not sufficient to produce the kindling temperature of the foil, and the bulb will not flash. Once a bulb has been flashed, it cannot be used again because the supply of oxygen and metal have been used up. Most modern flash bulbs are coated with a thin layer of a plastic material which reduces the possibility of explosions and shattered glass. It's recommended, however, that a clear plastic cover be placed over the flash bulb and reflector in order to afford added protection to the subjects being photographed.

Why are cyanides poisonous?
As we have already discussed, *enzymes* are protein molecules produced by the body and used to facilitate or speed up many of the chemical reactions going on in living cells. Without the help of enzymes, the chemical reactions in living cells would take place at a much slower rate, or not at all. It's believed that cyanides are poisonous because they combine with enzymes and prevent them from performing their essential functions.

While scientists have learned what enzymes do, they are still not quite sure about *how* they do it. There's a theory, however, that tries to get at the basis of enzyme action. It's based on the fact that enzymes seem unable to catalyze, or help along, more than one specific chemical reaction. In considering the problem, let's pursue the following line of reasoning. Let's imagine that we have a microscope powerful enough to reveal the innermost details of an enzyme molecule. As we look into the microscope, an enzyme might seem to consist of a large number of spherical objects. Each of these touches one or more of the others. The group of objects seems to form a more or less complicated mass of balls. Upon closer examination, these balls seem to vary considerably in size, but the entire molecule is composed of only a few different sizes. After some thought, we conclude that the smallest ones must be hydrogen atoms, which are very light in weight, while the larger ones must be carbon, nitrogen, oxygen, and possibly sulfur and phosphorus atoms. As we look closer, we find another group of atoms which looks exactly like the first; this must be another enzyme molecule. There are many such groups of atoms, and they all look exactly alike. Perhaps this is a key to the problem. We mentioned earlier that enzymes are used to help along certain specific chemical reactions. Let's assume that our particular enzyme is capable of facilitating the union of a water molecule and a molecule of an organic substance (one containing carbon) taken into the body as food. Let's place some of these "raw materials" for the enzyme in contact with it and see what happens. As we watch the proceedings under the microscope, we notice that a water molecule finds its way to a certain portion of the enzyme. As it reaches this point, it "pops" into the enzyme like a key fitting snugly into a lock. In much the same way, a molecule of the organic substance fits into the enzyme right next to the water molecule. While in this condition, the water molecule "grabs on" to the organic compound. In so doing, it distorts the enzyme enough to cause both compounds to be ejected from their "keyholes." From that time on, the water molecule and the organic compound remain joined together. The combination of the two is now a new molecule. As soon as the new molecule moves away from the enzyme, new "raw materials" move in to repeat the process.

Although this picture of enzyme action probably oversimplifies the actual situation, it is useful, nevertheless, in illustrating the kind of action going on. With this picture in mind, it's possible to understand how certain drugs and poisons work. We know, for example, that cyanides are poisonous because they combine with one of the essential enzymes of living cells. According to the theory given above, cyanides probably enter the enzyme because they are similar to one of the enzyme's "raw materials." Once in the "keyhole," however, they refuse to combine with the other "raw material" and so are never ejected. In this way, the enzyme becomes blocked by the cyanide, and finds it impossible to perform its normal function. Similarly, drugs of the sulfanilamide family are believed to have a shape that is similar to a "food" required by certain microbes. The drugs are enough like the microbes' normal food to trick the microorganisms into absorbing them, but enough different so that they do not behave as food does.

What was the "Mauve Decade"?
In 1856 William Perkin, a seventeen-year-old English schoolboy, was given the assignment of making quinine, an important drug. In the course of his chemical experiments, he noticed that an unexpected colored substance was formed. With the help of his instructor, Perkin

discovered that this colored substance was formed as a result of impurities present in the chemical, *aniline,* that he used as a starting material. Perkin wondered if this new substance could be useful as a dye. If so, he would be the first man to produce an effective dye synthetically. He later showed that his new compound, which came to be known as *mauve,* was an excellent dye and that others could be made in a similar manner. His discovery unlocked the secret of cheap dyes and brought the pleasure of royal colors within the grasp of everyone. So thoroughly did his discovery affect the life of his time, that some writers called the years following, the *Mauve Decade.*

Aniline, the substance that influenced Perkin's discovery, is made from *benzene,* which, in turn, comes from coal tar. Dyes made in this manner are called *coal tar dyes.* Scores of chemists are at work today making new coal tar dyes similar to those made by Perkin.

While coal tar is extremely useful in today's chemical industry, it was once nothing but a headache for gasmakers. In the manufacture of gas, coal is heated and four products are obtained: coal gas, coke, ammonia and coal tar. But the early gas producers had absolutely no use for coal tar. It was not yet used on roads; coal tar dyes were unknown; and the many other coal tar derivatives were yet to be discovered. Consequently, coal tar was disposed of in near-by rivers and streams, killing the fish they contained, and firmly establishing the odorous reputation of the gasworks. Luckily, chemistry found a solution to this unhappy problem. Today, coal tar can be made into over 200,000 compounds. Of these, many thousands are manufactured regularly in the form of dyes, medicines, explosives, plastics and chemicals. From a humble though colorful beginning, coal tar has found its place in our world through chemistry.

What causes the crescent moon?

The moon, unlike the sun, has no illuminating power of its own; it shines by reflecting sunlight. Since the moon is a sphere, only half of it can be illuminated by sunlight at any one time. The portion of the illuminated half that we see determines the "phases," or variations in the appearance of the moon—crescent, quarter, full, and so on. The ancient Babylonians believed that the moon had a "bright" side and a "dark" side and that it rotated slowly in the sky, thereby presenting to the earth all phases from full (bright side) moon to new (dark side) moon. Of course, this isn't the correct answer. The phases of the moon are actually caused by the relative positions of the sun, the moon and the observer. You can prove this to yourself in a darkened room with the help of a ball and a flashlight. Hold the ball in one hand and shine the light on it from various directions. Start with the ball between the flashlight and your eyes. Now rotate the flashlight until you see a thin crescent of light on the ball. A little experimentation will result in the production of all of the "phases" of your miniature "moon."

When we first see the new moon in the western sky, it appears as a crescent that gradually increases in size until it is "half-full." It is then said to be in the first quarter. It passes through the three-quarters full, or "gibbous," stage and finally the whole visible hemisphere is illuminated, when the moon is said to be "full." The earth is then roughly between the moon and the sun with the light of the sun reaching the moon from over the earth's shoulder, so to speak. From that point on, the opposite procedure sets in as the moon progresses toward the sun. It passes through the gibbous, last quarter and crescent stages until it moves between the earth and the sun, starting off as another "new" moon. During the first half of this cycle, when the illuminated portion is growing larger, the moon is said to be *waxing*. After it is full, it is said to be *waning*. But whether waxing or waning, the points of the moon's crescent always point *away from the sun*. You can amuse your friends by using this piece of information to determine the sun's direction when the moon is out. Or you might use it to check the crescent moons in the many paintings and drawings in which they are depicted. You will find that artists have painted pictures with the moon's crescent pointing in all conceivable directions, many of which are impossible of attainment in real life. It's impossible, for instance, to have a moonlit night in which the crescents point downward. Such a condition could only exist with the sun above the horizon—a pretty good definition of daytime!

What keeps an artificial earth satellite up in the sky?

Scientists tell us that the old adage, "Everything that goes up must come down," isn't necessarily so. If we fire a projectile into the sky it will probably return to earth under the influence of the force of gravity, but this is only because we didn't shoot it fast enough. If a muzzle velocity of about 24,000 miles per hour is attained, the projectile will leave the earth behind and travel into outer space, never to return. This velocity is called the *escape velocity*—the velocity a projectile must attain before it can escape from the gravitational pull of the earth.

Another interesting velocity is that required by a satellite which is to circle about the earth. If a bullet is fired horizontally, it will eventually fall to earth due to gravity. This, again, is because of insufficient speed on the part of the projectile. Let's examine this situation a little more closely. The earth, of course, has a curved surface. A bullet (or satellite) shot horizontally would keep traveling horizontally were it not for gravity. It would eventually leave the earth and travel in a straight line indefinitely. But we can't neglect gravity and its effect on the bullet. Gravity causes the path of the bullet to curve toward the earth. The amount of this curvature depends on the speed of the bullet. If it is shot out fast enough, its horizontal motion (neglecting air friction) can compensate for the amount it falls, thereby keeping it a constant distance from the earth's curved surface. For relatively low heights above sea level, this speed is about 17,000 miles per hour. The required velocity decreases as the height of the satellite increases. Close to the earth, such a satellite takes about 1½ hours to circle the earth. At a distance of 22,300 miles, the satellite travels at about 7,000 mph and requires just 24 hours to circle the earth. Since this is the time required for the earth to make one revolution, the satellite would remain permanently over the same spot on the earth's surface, providing a convenient terminal for interplanetary travel. Sometime after July 1, 1957, the U.S. Government will launch an artificial satellite in accordance with these principles. It will zoom 300 miles into the atmosphere carrying a 20-inch golden ball. When the ball is released by the rocket, it will circle the earth in an elliptical path. At its closest advance, it will be about 200 miles away; its most distant excursion will carry it about 1,500 miles from the earth. It will take 90 minutes to complete a single journey around our planet at an average speed of 18,000 mph. Crammed with electronic gear, the golden ball will be capable of relaying valuable data to ground stations concerning cosmic radiation and other phenomena which are almost completely unknown at the present time. Project Vanguard, as the satellite program is known, is part of the United States's participation in the International Geophysical Year, July, 1957 to

December, 1958. Scientists from forty countries are involved in the many aspects of this program. Project Vanguard is under the control of the Naval Research Laboratory. The satellite will continue to broadcast messages to the earth for a period of two weeks. We can only guess, however, how long it will continue to circle the earth— perhaps a month, perhaps a year.

Why do metals expand when heated?

Metals, like all matter, are made up of molecules which are in continual motion. Although molecules of a solid do not move about freely, they vibrate back and forth much like a cork bobbing up and down on the waves. When a metal is heated, the molecules vibrate at a faster rate, their speed depending upon the temperature at any given time. As the molecules speed up, they strike their neighbors harder and at an increasing rate. This tends to push the molecules farther apart, and the substance takes up more room. For this reason, most things tend to expand when heated and contract when cooled.

What was "the year without a summer"?

The year 1816 was one of unbelievable weather for the inhabitants of New England and eastern Canada. It was known as the "poverty year," or the "year without a summer." Winter weather just refused to depart and temperatures stayed low throughout the summer. Although the entire world was plagued with cold spells, the northeastern part of North America was hit most severely. An early June storm killed early crops and blanketed the zone in deep snow. Many summer birds and even some farm animals froze to death. The entire summer was marked with frosts and snowfalls, turning what should have been summer into the most unusual kind of winter. Many late crops were destroyed by the unexpected weather. Of those that survived, many did not mature.

What caused this climatic nightmare? Scientists of the day were inclined to blame it on sunspots which, for the first time in recorded history, were visible to the unaided eye. While sunspots may have had something to do with the problem, it is suspected that the unusual weather was connected with an unusually severe volcanic eruption. In the spring of the previous year, a volcano on the opposite side of the earth had erupted with tremendous violence. The rumbling mountain continued to give off unbelievably large quantities of smoke and ash for five days. A gigantic cloud formed over the area. This black cloud was dense enough to dim the sun for hundreds of miles around. It is theorized that this dust was carried by winds to all parts of the earth. Perhaps it was particularly dense over the area of North America in question. In any event, it is suspected that this dust reflected an abnormal amount of the sun's heat back into space. This would account for the lower temperatures prevailing in that section even though the calendar called it summer.

Proponents of this theory believe that most of the dust had settled with the help of rain and snow before the next summer arrived. The temperatures existing at that time, therefore, would be more nearly normal.

It's interesting to find that some climatologists have a theory that runs counter to our most basic belief about the weather. If asked how the earth's climate might get warmer, most of us would vote for an increase in the sun's radiation. Curiously enough, some scientists believe that just the opposite would happen! If the sun were to give off more heat, they believe, the resulting clouds would reflect a greater proportion of the sun's energy back into space. This indirect effect would cause the earth to get colder; it might even result in an ice age!

What are air masses and fronts?
Large bodies of air can usually be found traveling from west to east across the United States. On occasion, one of these may remain for a few days over the same region before passing on. Under these conditions, the body of air assumes a temperature and humidity corresponding to the land areas beneath. A body of air of this kind is called an *air mass*. Air masses formed in the far north are called Arctic air masses, while those formed near the Equator are called tropical air masses. Whenever air masses from different regions meet, a boundary surface is formed between them. This is called a *frontal surface*. This surface is invisible and uneven, changing constantly as one air mass presses against the other. We might expect a frontal surface to be a more or less vertical boundary separating the two air masses, but such is not the case. Since cold air is more dense than warm air, we always find that the cold air mass "hugs" the ground while slipping in under the warm air mass. This causes the frontal surface to stretch out in a gradual slope. It may not rise vertically

much more than one mile in three hundred miles of horizontal distance. The warmer, less dense air is always on top of the frontal surface while the cold air is always found beneath it.

The bottom of a frontal surface, the part in contact with the ground, is called a *front*. As with its frontal surface, a front shifts constantly as parts of the advancing air mass change their directions. Its general progress, however, is from west to east across the continent. A *warm front* is the front of a warm air mass that is advancing on a colder air mass. Since warm air is less dense than cold air, it cannot crowd out the cold air mass as fast as it would like to. Instead, some of the advancing warm air passes over the cold air as if waiting for

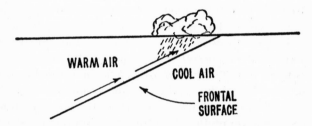

FIG. 29. ONE CAUSE OF PRECIPITATION

When a warm air mass comes in contact with a cold air mass, some of the warm moist air is forced to rise. This lowers the temperature of such air, sometimes resulting in precpitation.

it to move out. The frontal surface slants up and over the cold air mass. As a result, a warm front is usually hundreds of miles behind the most advanced portion of the warm air mass that caused it. In reverse fashion, the face of an advancing cold air mass slants backward. The cold air mass advances like a wedge, hugging the ground, and sliding under the warmer air. This forces the warm air upward.

The approach of either a warm front or a cold front forces warm air to rise. Since this reduces the temperature of the warm air so affected, it is possible that precipitation may result. Warm air usually contains a considerable amount of moisture. When such air cools, a temperature is reached where it can no longer hold so much moisture, and precipitation is the result.

Sometimes two air masses come together under conditions that are not favorable to cause either to advance. The front that results is called a *stationary front*. On other occasions, a cold air mass may overtake a warm air mass, lifting all of the warm air upward. The boundary line in such a situation is called an *occluded front*. An occluded front also results when two cold air masses advance on a warm air mass. The warm air is then squeezed and pushed upward as the cold air masses converge on it.

What causes ptomaine poisoning?

The ptomaines are substances produced by the decomposition of proteins and they are nonpoisonous. The expression, "ptomaine poisoning," therefore, is a misnomer. The sudden and acute intestinal illnesses usually referred to as ptomaine poisoning are the result of various bacteria that enter food and produce substances that *are* extremely poisonous. The microbes generally responsible for such poisoning include *Clostridium botulinum,* and certain strains of *streptococci* and *staphylococci.* The former produces an exceedingly poisonous toxin in the absence of oxygen, as in a sealed can, and is by far the most dangerous of all of the food-poisoning bacteria. It's normally found in the soil and sometimes in the intestinal tract but in such cases no harm results. But if the organism is not destroyed in the canning of food, it will multiply and produce its deadly poison, botulinum toxin. Poisoning by this toxin has become rare from commercially prepared foods because of the precautions taken by food processers. The standards used in home canning are another matter, however. In preserving nonacid foods such as peas, corn, beans and meat, great care must be taken to kill all of the bacteria, and the spores from which they germinate. This can be done by sterilizing foods at the high temperature of steam under pressure. No precaution is too great to be taken in preparing such foods. Unfortunately, it's not always possible to identify food that is contaminated with botulinum toxin. It may have no unusual smell, *and a single taste has been known to be fatal.* The only safe way to protect yourself against this poison is to process food well enough to kill the organism; to boil all home canned food for at least twenty minutes before tasting; and to throw away without tasting any food that has changed in appearance or color.

The usual symptoms of botulism include vomiting, diarrhea and difficulty in swallowing. The toxin involved in this kind of food poisoning affects the nerve endings and causes paralysis. Death is due to the paralysis of the respiratory system and other vital functions.

Food poisoning caused by the streptococci and staphylococci organisms mentioned earlier are disturbing, but not too serious. When creamed foods, custards and salads are allowed to stand at room temperature for some time before serving, the bacteria may enter them and multiply greatly. This results in the production of large amounts of toxins which cause the illness upon being eaten. This kind of poisoning can be avoided by refrigerating food before and after handling and by eating uncooked foods promptly after handling.

Has anyone ever seen a molecule?

Most molecules are extremely small and for this reason have long eluded the eye of would-be observers. Recently, however, photographs of molecules have been taken with the aid of the electron microscope. Photographs of large virus molecules show them to resemble clusters of minute baseballs, about a millionth of an inch in diameter. In contrast, an atom of hydrogen has a diameter of about two-billionths of an inch. Recently, a new technique has been used to photograph atoms. It's done with a new type of microscope that uses both light and x-rays and that magnifies about 2,200,000 times. Iron sulfide was the first compound to be photographed by this method. The photograph shows the compound to consist of an iron atom, flanked on either side by smaller sulfur atoms.

Before we leave the subject, there are some other questions about molecules that keep popping up. How fast do they move? The average molecule of air moves at a speed of twelve hundred miles per hour—much faster than the speed of sound. How far apart are they? The molecules of most solids are about one one-hundred-millionths of an inch apart. Do they collide with one another? Yes, there are about ten thousand trillion trillion collisions per second in one teaspoonful of air! How plentiful are they? There are about seven million million billion molecules in a glassful of air. Perhaps we should close with the often repeated story of the physics instructor found sobbing at his microscope. When asked about his sadness he replied, "Molecules are so awfully small."

Why is snow white?

Since snow is nothing more than frozen water, it should be as colorless as ice. Its whiteness is due to the reflection of light from the many surfaces of the large number of ice crystals in each snowflake. Its opaqueness is a result of the presence of air in the snowflake. Each snowflake consists of a large number of minute but definite forms known as crystals. Many substances, like water, form crystals when they freeze or harden. In all such crystalline substances, the molecules come together in a special configuration which is peculiar to the substance involved. Such a configuration, or geometrical form, is called a *crystal*.

When water vapor in the atmosphere freezes, it forms clear transparent crystals that are so small as to be invisible. These are carried up and down in the atmosphere, alternately falling for a while and then rising under the influence of varying currents of air. In so doing, the crystals form nuclei which develop in size, gathering together in families of a hundred or more. When a family of ice crystals is large enough, it slowly floats down toward the ground and we call it a snowflake. Snowflakes vary in size from over an inch in diameter down to the tiniest of specks. The actual size depends on atmospheric conditions. As you have probably noticed, the largest ones form when the temperature is close to 32° F., while colder temperatures result in smaller ones.

The crystals that constitute a snowflake are always arranged about a center at an angle of either 60° or 120°. They invariably form either six-pointed stars or thin plates of hexagonal shape. From the center of the six-pointed stars, branches spread out in six directions, each branch being identical to the others. In spite of this sixfold repetition of form within a snowflake, no two snowflakes are ever found alike. Although individual snowflakes are altogether different from each other, the branches of any single snowflake will always be identical. They exhibit extremely beautiful forms that have often been used by craftsmen as the basis of their designs.

While snow is usually white, there have been several instances on record in which colored snow has fallen. Darwin wrote that during his passage of the Cordilleras, in 1835, the footsteps of the mules were stained red as if the hoofs of the mules were covered with blood. It was later found that the snow contained countless minute plants called *algae*. In this case the variety was *Protococci nivales,* consisting of microscopic spheres measuring less than one-thousandth of an inch in diameter. These plants are so prolific that an Arctic or Alpine landscape has been known to change from white to red overnight as a result of their presence. Similarly, another species known as *Spaerellae nivales* produces drifts of rose-colored snow.

Are there any "new" volcanoes?

On February 20, 1943, a new volcano was born in a Mexican corn-field. You can imagine the astonishment of the farmer as he stopped his plowing to investigate a thin column of smoke rising from the middle of his field. The smoke wasn't coming from a fire on the surface, but was rising out of a small hole in the ground! After put-ting a stone over the hole to put out the fire, he returned to his plowing. Later, however, he noticed that even more smoke was coming out of the ground. Now quite alarmed, he began to recall the ground tremors that had occurred recently, and the fact that the ground felt warmer to his bare feet than it ever had before. Realizing that the problem was too great for him to cope with, he went to the mayor of the near-by town to tell about the unusual fire on his property. After the normal amount of skeptical indecision, a party went to the farm to find out what it was all about. When they arrived, they found dense clouds of smoke billowing out of a hole about thirty feet deep. That same night, the first eruption occurred, sending smoke, cinders and volcanic matter a mile into the sky. Every few seconds a new explosion took place as great quantities of volcanic rock were blown from the crater's mouth. Sometimes these projectiles would explode high in the sky, adding their bit to the fury within the earth. Lightning flashed every minute or so, illumi-nating nature's great display of destruction. Two days after the first eruption, lava began to flow. The volcano, named Parícutin after the first town destroyed, continued to erupt for many months. Lava and other volcanic matter destroyed farms, forests and villages. Farming within fifty miles of the new volcano had been made im-possible. Of course, the birth of Parícutin was a great disaster to the people involved. To compensate to some extent, however, it has given scientists, for the first time in recorded history, an example of a volcano that can be studied from its very beginning. Perhaps the knowledge gained from it by geologists will help to reduce future devastation caused by other volcanoes.

How can birds stand on electric power lines without getting electrocuted?

An electric current must travel in a closed loop. Take the ordinary battery, for example. It must have two terminals in order to provide electric power. When these terminals are connected to a lamp, a complete electrical circuit is formed. Electrons travel from the negative post of the battery, through the tungsten wire of the lamp, and back to the positive post of the battery. If any part of the circuit becomes disconnected, the current stops flowing and the lamp goes out. The same principle protects birds perched on high-voltage power lines. Power is transmitted from the generating station to our homes by a pair of wires: one wire carries electrons from the generator to the electrical appliance; the other wire returns the electrons to the generator. Before a bird can be electrocuted, it must touch *both* of these wires, or one wire and the earth. Most power systems are connected to the earth in some manner so it is usually fatal for a "grounded" individual to touch even *one* power line. If he should, the current passes from the wire, through his body, and back to the generator through the ground. In either case, a bird's small size makes it impossible for him to touch two wires at the same time, or one wire and ground. This makes it perfectly safe for birds to perch on power lines.

Why do airplanes cause flutter in television pictures?

Radio waves normally travel in a straight line from the broadcasting station to the television receiver in your home. But the wave lengths used in television broadcasting show a marked tendency to be reflected from metallic objects like airplanes. When an airplane is overhead, some of the radio energy from the television station is reflected down into your TV antenna. In this way, two signals enter your set: a desired signal that comes directly from the station and an incidental signal that bounces off the airplane. This in itself wouldn't be too great a problem were it not for the timing involved. Radio waves travel extremely fast—at the speed of light, in fact—but they do not travel instantaneously. It takes a short but measureable length of time for the signal to travel from the station to your TV set. Naturally, the signal that travels the most direct route gets there first. The incidental signal that is reflected from the airplane travels a longer route and arrives at your set a fraction of a second later. Depending upon the exact time difference, the two signals will either reinforce or cancel one another. But the airplane is in motion. This means that the length of the path that the reflected wave must travel will change from one instant to the next. In other words, the distance from the station to the plane to the TV set will change constantly, depending upon the location of the plane. This results in a continually changing time lag between the reflected signal and the direct signal. When the time difference is advantageous, the signals add and the picture gets darker; when the time difference is disadvantageous, the signals subtract and the picture gets lighter. This produces a fluttering picture as the signals alternately reinforce and cancel each other in rapid succession.

While we're on the subject, someone will want to know about "ghosts." A clear, sharp television picture results when a single strong signal is received. But sometimes a "ghost" shows up on the screen, lighter than the desired picture and displaced somewhat to the side. In some cases there may be several ghosts, one beside the next. Since ghosts are usually lighter than the desired picture, they are somewhat "transparent," allowing the viewer to see another picture "through" them. This probably gave the phenomenon its name. Ghosts of this sort result from the reception of two or more signals which arrive by different paths. As was mentioned earlier, the desired signal takes the most direct path from station to TV set. But a near-by object, such as a water tower or large building, may reflect another signal from the same station into your set. This signal will arrive a fraction of a second later than the desired signal because of the longer path involved. This results in the reception of a second picture (a ghost) that is displaced somewhat to the right of the desired picture.

What is the law of gravitation?
By the year 1666, Newton had satisfactorily explained the forces that account for the observed motion of heavenly bodies. He theorized that the (gravitational) force of attraction between these bodies *varies inversely as the square of the distance separating them.*

Let's use an imaginary example to illustrate this statement. Suppose that two bodies are located a certain distance apart in space. In addition, imagine that they are far enough away from other bodies so that no other forces act on them. Under these conditions, we measure the force of attraction between the bodies and find it to be 1,000 pounds. Now let's move them twice as far apart and measure the force again. When we do this, we find that the force is only 250 pounds—one-fourth as great as before. Let's see if this agrees with Newton's principle. We multiplied the distance separating the objects by 2. If we square the number 2 we get 4. But Newton's law is an *inverse* law. This means that the force goes down as the distance gets greater. Then we must divide the original force by 4 (2 squared) to get the new force. One thousand pounds divided by 4 equals 250, so our hypothetical problem agrees with Newton's *inverse square law* given above.

But this is only part of the law of gravitation. It would seem to be only a scientific hop, skip and jump to a full statement of the law of gravitation, but instead Newton delayed almost twenty years in announcing it. Was this delay a result of his shyness and disinclination to engage in controversy? It's probable that the real reason lay in quite another direction. The gravitational pull of the earth on any object, whether a battleship or the legendary apple that supposedly fell on Newton's head, depends on the accumulated action of many forces all acting in concert. Specifically, we can't say that the earth attracts an apple; we must say instead that every minute particle of the earth attracts every minute particle of the apple. The effective force which pulls the apple to earth is the resultant of all of these multitudinous and microscopic forces. You can see what a hopelessly complex problem this must have presented to Newton. There just wasn't any form of mathematics available to handle such a problem. To find the solution, Newton had to invent a new mathematical tool, the *integral calculus*. With it he was able to satisfy his scientific conscience and make known the law of gravitation in the third volume of his great *Principia*. The law states that the gravitational attraction between two objects varies in direct proportion to the product of their masses, and in inverse proportion to the square of the distance separating them. In simplified terms, the force is greater for heavier bodies, and smaller for lighter ones. It increases when the bodies come closer together, and it decreases when they separate.

How does a thermostat regulate temperature?
It's surprising to discover the extent to which the ordinary thermostat has become a part of modern living. We use it to control the temperature of our homes, our refrigerators, our electric irons and

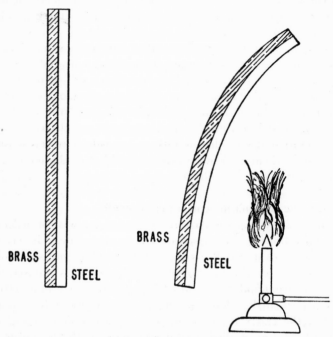

FIG. 30.   PRINCIPLE OF THE THERMOSTAT
A bimetallic bar is made of two dissimilar metals welded side by side. When such a bar is heated, it bends in an arc of a circle.

deep fryers, our electric stoves and even the cooling systems in our automobiles. A thermostat can be adjusted so that it will turn on an oil burner when the room temperature falls below a certain reading and so that it will turn it off when the temperature rises above another reading. Wherever there is need for close temperature control, the thermostat comes to our rescue.

The essential part of most thermostats is a *bimetallic bar* made by welding strips of two dissimilar metals side by side. The metals commonly used are steel and brass. This results in a bar that is brass on one side and steel on the other. When we heat such a bimetallic bar, we find that it bends in an arc of a circle. This is a result of the different rates of expansion of the two metals. Brass expands almost twice as much as steel when both metals are heated the same amount. This causes the brass side of the bar to elongate more than the steel side. The only way that this can happen is for the bar to bend and form a circular arc.

This movement makes a bimetallic bar quite suitable as the temperature-sensitive element in a thermostat. One end of the bar is fixed and the other allowed to move as a result of temperature changes. An electrical contact is placed on the movable end and another on a near-by fixed location. In this way, the bending of the bimetallic strip is caused to make or break the electrical circuit. Through proper design, thermostats can be made extremely sensitive. This enables them to provide accurate control of the temperature of our many electrical appliances.

How was the Arizona meteor crater formed?

Scientists believe that the Arizona meteor crater was formed by an explosion following the impact of a large meteorite. The crater is about two-thirds of a mile across, and its floor is about six hundred feet below the level of the surrounding terrain. It is believed to be about fifty thousand years old. The meteorite that produced the crater is estimated to have had a diameter of about two hundred feet. But the crater itself would require two billion tons of rock to fill it! The great disparity in size between the meteorite and crater makes it quite evident that a tremendous explosion must have followed the impact of the meteorite with the desert floor. It is believed that a great quantity of gas was suddenly formed by the friction between meteorite and earth. As the meteorite drilled deep below the surface, a temperature of millions of degrees centigrade was reached which produced an amount of gas perhaps equal in weight to the meteorite itself. The sudden release of this gas produced an explosion that emptied the surrounding area of two billion tons of rock. The remains of the meteorite are believed to rest beneath the south rim of the crater, at a depth of about fifteen hundred feet. Early attempts to dig for the meteorite failed because it was assumed that its path would have carried it directly below the center of the crater.

The craters of the moon are also believed to be of meteoric origin. Studies of the moon show that it has literally thousands of such craters. There seems to be little doubt that they came into existence as a result of countless collisions with meteorites during the four billion years or so that the moon has circled the earth. Since the moon has no atmosphere, there has been no erosion to wear away the lunar craters. While those on earth have been continually subjected to the effects of weather, those on the moon remain unchanged from age to age. A few million years would completely obliterate a moderate-sized crater on the earth. This explains why we have so few of them, while the moon has so many.

A meteorite capable of forming the Arizona crater would completely devastate a much larger area in a flash. A relatively small meteorite estimated at a few hundred tons, fell in a remote part of Siberia on June 30, 1908. The blast was felt and heard four hundred miles away! When the region was explored in 1927, it was found that an area at least forty miles in diameter had been completely demolished. Within this area, trees had been knocked over and stripped of bark and branches. Luckily, such meteorites are relatively rare, and the chances of their striking a highly populated portion of the earth are reassuringly remote.

How do "smudge pots" protect orchards from freezing?

The smudge pot is the most common device used to protect citrus-fruit crops from the effects of freezing temperatures. It consists of a torchlike mechanism that produces dense smoke from burning oil. It's not the heat of burning oil that is effective in this application, but rather the insulating qualities of the smoke itself. The pots are usually mounted on short poles and distributed throughout the orchard. When the Weather Bureau predicts frost, the pots are lighted. Dense clouds of smoke from the burning oil form a heavy blanket that covers the entire orchard. This blanket of smoke acts as an insulator, reducing the amount of heat that can escape from the ground and from the air surrounding the trees. In addition, the smoke manages to extract heat from the atmosphere itself. As air cools below its dew point, the water vapor that it contains condenses upon the smoke particles. As it condenses, it must give up its heat of vaporization. This is the same heat that the moisture had to absorb when it changed from water to water vapor. This heat is added to the heat that the smoke blanket prevents from escaping from the area. In this way, smudge pots are able to keep the air temperature above freezing and reduce the possibility of damage to the crop.

Another method used in protecting crops from freezing makes use of infrared (heat) rays. It consists of a special oil-burning heater equipped with a reflector that sends heat rays in all directions among the plants. It is used not only in orchards, but also in fields, gardens and greenhouses.

Perhaps the most unusual method of crop protection makes use of the fact that air thirty to forty feet above the ground is usually five to ten degrees warmer than the air just above the ground. Rotating wind machines are installed on towers thirty-two feet high to use this warmer air in preventing frost. Every few minutes, a gust of warmer air is blown among the trees of the orchard. If necessary, additional heat is provided by oil heaters distributed among the trees. This method of frost prevention is claimed to be the least expensive of all.

What is the even-tempered musical scale?

The notes that make up the musical scale bear a fixed relationship to one another. If the lowest note of the scale—C, for example—has a frequency of 256 vibrations per second, the next-higher note will have a frequency that is 9/8 x 256, or 288 vibrations per second. Similarly, the next one up the scale will be 5/4 x 256 in frequency. Expressed in the form of a mathematical ratio, the full scale can be given as 1:9/8:5/4:4/3:3/2:5/3:15/8:2. If we multiply each of these numbers (or fractions) by 256, we get the diatonic scale for the key of C. This scale is the one used by physicists.

C	D	E	F	G	A	B	C	D	
256	288	320	341.3	384	426.6	480	512		Key of C
	288	324	360	384	432	480	540	576	Key of D

The top row of frequencies represents the scale in the key of C. The frequency of D is 288 vibrations per second which is 9/8 times 256. Each of the other numbers of the top row is obtained in the same way by multiplying 256 by the appropriate number from the ratio given above. Now let's see what happens when we calculate the diatonic scale for the key of D. To do this, we multiply 288 (the second note in the C scale) by each of the numbers from the ratio. This gives us the bottom row of numbers which constitutes the scale in the key of D. Four of these notes are different from those obtained for the key of C. Note particularly how far the notes F and C are from the same notes in the key of C. This change in key necessitated the introduction of several new notes to the instrument. If we were to continue this process with all of the other possible changes in key, we would find the need for still more notes. To give each key its proper equipment of notes would require about 70 notes to one octave. To avoid this complicated arrangement, a compromise has been arrived at whereby each octave is made up of 13 half-tones equally spaced. This scale is called the even-tempered scale. Each half-tone has a frequency that is about 1.06 times the preceding note. While most of the notes are not exactly right in frequency, the departure from an ideal scale is slight and most of us are not disturbed by it.

Since the notes in any scale bear a fixed ratio to each other, any frequency could be chosen as a standard, or starting point. Physicists have selected the standard of middle C = 256 vibrations per second. The standard for International Pitch is A = 435. Most concert orchestras now use A = 440. The even-tempered scale in International Pitch is as follows:

C	258.6	F#	365.7
C#	274.0	G	387.5
D	290.3	G#	410.5
D#	307.5	A	435.0
E	325.8	A#	460.8
F	345.2	B	488.2
		C	517.3

What is the water table?

When rain falls to the ground, it seeps into the cracks and pore spaces of the soil until it reaches a material through which it cannot pass. This may be an impervious layer of clay or granite, several hundred feet below the surface. Since water can't sink any further, it accumulates above the layer of impervious material. This results in a layer of soil and rock that is saturated with water. You might thınk of this saturated layer as an underground "lake" or slow‑moving "river." The water in this underground reservoir is called *ground water,* and its upper limit is called the *water table.* Above the water table, the pore spaces of the soil are filled mostly with air instead of water. The depth of the water table below the surface of the earth depends primarily upon the location of an impervious layer. Of course, local rainfall and terrain are also important. In dry periods, the water table drops as water is lost to evaporation, plants and a slow flow to swamps and lakes. In rainy weather, the water table rises as rain is added to the supply of ground water. In hilly country, the water table may be higher than the level of the land in some places. This causes the water to come up out of the ground in search of its own level. If the low spot has no natural outlet, a lake is formed.

Is there a limit to how cold something can get?
In his study of gases, Jacques Charles showed that the volume, pressure and temperature of gases are all interrelated. If he changed one of these factors, one or both of the others would change. To make this clear, let's imagine that we have a cylinder full of air. The cylinder is closed at one end and has a movable piston at the other. By moving the piston up and down we are able to change the pressure of the air in the cylinder. Let's further imagine that the pressure inside the cylinder is 15 pounds per square inch, and that the temperature of the air is 0° C. Now let's cool the air sample within the cylinder to −1° C. When this is accomplished, we find that the pressure within the cylinder is less than the 15 pounds per square inch that we started with. In order to regain the original pressure, we must compress the air within the cylinder slightly. After we make the adjustment by pushing the piston into the cylinder a bit, we find that the volume of the sample of air has been reduced by 1/273rd of its original volume. In other words, a one‑degree reduction in temperature required a volume reduction of 1/273rd to keep the pressure unchanged. If we reduce the temperature another degree, we find that we must push the piston in exactly the same amount to keep the pressure constant. Each time we lower the temperature one degree, the volume must be reduced by 1/273rd of the volume at 0° C.

You might reason that if this process were carried out long enough, all of the gas would be gone (that is, it would have zero volume) at −273° C.! Of course, this can't happen because all gases change into liquids before this low temperature is reached. But Lord Kelvin used similar reasoning to develop a theory concerning the lowest possible temperature. He called the temperature at which gases *theoretically* cease to have a volume, *absolute zero*. He believed that at this temperature all molecular motion stops and gases cease to exert pressure. This temperature is −273° C. The gas used in Navy dirigibles, helium, liquefies at −268.9° C. and freezes at somewhat less than −272.0° C. Temperatures have been obtained in the laboratory as low as a few tenths of a degree above absolute zero. This leads most scientists to believe that −273° C. is about as cold as anything can get. If you're wondering about the other end of the temperature scale, there doesn't seem to be any limit to how hot a substance can get.

INDEX

Raindrop, electrical charge on, 18
  shape of, 5
Rainfall, and tree rings, 168
Rattlesnakes, seeing in the dark, 112
Reactor, atomic, principle, 100
Recapitulation in embryos, 97
Recognition of faces and names, 137
Reconditioning, as a control of fear,
  108
Recorder, tape, 43
Records, phonograph, 90
Redwoods, height of, 181
Reflection, internal, 103
Reflex, conditioned, 34
Reflex actions, as opposed to thought,
  42
  in wasps, 32
Refraction, in lenses, 86
Refrigerator, and cooling the kitchen,
  172
  location of the freezing compartment
  of, 163
Relative humidity, and cloud forma-
  tion, 262
Relativity, theory of, 70
Remembering of faces and names, 137
Rennet, in cheese-making, 251
Resins, and strength of trees, 174
Resonance, in television tuner, 13
Retina, 97, 113
Reverberation, 153
Rhesus monkey, Rh factor, 51
Rheumatism, 110
Rh factor, 51
Rickets, 156
Rifle bullet, reason for spin, 154
Ringworm, 236
Rio Tinto, colors of, 249
Rivers, characteristic colors of, 248
Rock, igneous, 126
  impermeable, and artesian wells, 165
  manufacture of, by oceans, 83, 163
  plastic flow and earthquakes, 44
  sedimentary, 84
Rocket ships, in a vacuum, 1
Rods, use in eye, 62
Roller coasters, 37
Root pressure, 39
Roots, direction of growth, 221
Rubber, elasticity of, 92
  first use of, 257
  naming of, 257
Ruby, 64
Rust, control of, 83
Rutherford, splitting of the atom, 160

Sailboat, tacking of, 117
Salt, and ice mixture, low temperature of, 24

Salt—*Continued*
  caking of, in humid weather, 16
  in salt lakes, 141
  radioactive, 73
Sanctorius, 78
Sand, formation of, 11, 126
Sand bars, formation of, 90
Sandstone, 84
Sap, rise of, in trees, 38
Sapphire, composition of, 64
Saprophytic fungi, 235
Sapwood, 174
Sargasso Sea, breeding place of eels, 163
  cause of, 248
Sargassum, 247
Satellite, artificial, 275
Scale insects, 96
Schiaparelli, Giovanni, 210
Schmidt, Johannes, 164
Schulert, Dr. A. R., 233
Sea animals, sleeping in water, 177
Seacoast, sinking of, 44
Seasons, of the earth, 79
Sea worms, longest, 228
  edibility of, 209
Seesaw, as a lever, 234
Senses, indifference to temperature, 131
Sequoias, age of, 204
  height of, 181
Sex hormones, in ulcer treatment, 125
Shale, formation of, 84
Shellac, 96
Shellfish, and limestone, 84
Ships, steel, ability to float, 104
  wrecked, depth of sinking, 50
Shower curtain, inward motion of, 257
Shrew, venomous glands in, 180
Sickness, motion, 20
Sight, of insects, 179
  of birds, 226
Silica, 155
  in diatomaceous earth, 248
Silicon, in atomic battery, 40
Silk, spider, 226
Silkworm, disease in, 203
Silt, in formation of rock, 83
Silver, plating of, 240
Silver sulfide, 135
Silverware, tarnished, cleaning of, 135
Singing in the shower, 153
Siphon, 127
Sirius, 90
Skin area, and food requirements of
  animals, 70
Sky, blue color of, 20
Slavery, among insects, 189
Smell, effect of, on taste, 25
  sense of, in fish, 117

305

Symbiosis—*Continued*
  in termites, 200
  in the lichen, 237

Tape recorder, 43
Tapeworms, 71
Taste, 24
  in insects, 178
Teardrop, shape of, 5
Telephone, 228
Television set, airplane flutter in, 284
  tuning of, 13
Temperature, absolute zero, 293
  and eel migration, 164
  control of, by thermostats, 287
  critical, of gases, 138
  effect of, on tuning of musical instruments, 133
  indifference of senses to, 131
  man-made, highest sustained, 152
  of the earth, hottest and coldest spots, 271
  of the hydrogen bomb, 86
  of a salt and ice mixture, 24
  upper limit of, 293
  yearly average in the U.S., 150
Termites, digestion of wood by, 200
  length of life, 223
Tetraethyl lead, 41
Theory of relativity, 70
Thermometer, clinical, 98
Thermoplastics, 122
Thermos bottles, nature of, 190
Thermosetting plastic, 122
Thermostat, 287
Thought, in insects, 42
Thrush, egg-laying habits, 95
Thunder, 49
Thyroxin, 181
Tidal waves, 151
  and the Krakatoa eruption, 110
  speed of, 110
Tides, 17
Time, sense of, in the palolo worm, 209
Time zones, 195
Tires, automobile, pressure rise in, 128
Titanium, in precious stones, 64
TNT, 138
Tobacco mosaic virus, 66
Tools, prehistoric, 223
Toothaches, sweets as a cause of, 35
Tortoise, life expectancy of, 223
Touch, sense of, in insects, 179
Toxins, 265
Tracks, railroad, length of, 119
Trade winds, and ocean currents, 179
Transformers, electrical, 238
Transpiration, of plants, 39
Tree, man-eating, 165

Tree rings, story of, 168
Trees, age of, 168, 204
  growth of, 174
  tallest, 181
Trial and error learning, 34
Trichina roundworm, 12
Trichinosis, 12
Trotting, 187
Trumpet, tonal quality, 44
Tsunami, 151
Tsuzuki, Dr. Masao, 234
Tuning, change of, in musical instruments, 133
Tunnels, ventilation of, 115
Turquoise, 64
Tyrothricin, 6

Ulcers, a man's disease, 125
Ultrasonic sound, 106
Underwater objects, location of, by sonar, 106
Universe, expansion of, 45
  formation of, 252
Uranium, and the age of the earth, 62
  and the atomic bomb, 47
Uranium-235, in the atomic reactor, 100
Urea, synthesis of, 170
Uric acid, and gout, 148
Urquhart, Dr. F., 220

Vacuum, and falling objects, 57
  flow of electricity in, 125
  transmission of sound in, 87
Vacuum bottles, 190
Vacuum tubes, 68
Valentine, Basel, 10
Valve, as a vacuum tube, 68
Vanguard, project, 276
Velocity, terminal, of falling objects, 57
Venomous mammals, 180
Venus's-flytrap, 165
Vigula divina, 9
Violin strings, bowing near one end of, 239
Virus, 65
  hydrophobia, 84
  in insects, 204
Visible spectrum, of light, 178, 261
Vision, duration of, 91
Vitamin, derivation of, 80
Vitamin D, 155
Vocal chords, use of, in speech, 85
Voice, human, 85
Volatility, 93
Volcano, newest, 283
Volcanoes, active, 95
  cause of, 95